APPLIED N = 1 SUPERGRAVITY

THE ICTP SERIES IN THEORETICAL PHYSICS – VOLUME 1

APPLIED N = 1 SUPERGRAVITY

Pran Nath†
Department of Physics
Northeastern University
Boston, Massachusetts 02115

Richard Arnowitt††
Lyman Laboratory of Physics
Harvard University
Cambridge, Massachusetts 02138

and

A. H. Chamseddine
Department of Physics
Northeastern University
Boston, Massachusetts 02115

† Lectures at the SUMMER WORKSHOP ON PARTICLE PHYSICS, 20 June – 29 July 1983, INTERNATIONAL CENTRE FOR THEORETICAL PHYSICS, Trieste, Italy

†† On sabbatical from Northeastern University

World Scientific

Published by

World Scientific Publishing Co Pte Ltd.
P O Box 128
Farrer Road
Singapore 9128

ISBN 9971-966-48-4
9971-966-49-2 pbk

Printed in Singapore by Richard-Clay (S.E. Asia) Pte. Ltd.

FOREWORD

Each summer from June through July, there are held at the Trieste Centre extended workshops on particle physics and condensed matter physics on areas of topical interest. The format is by and large informal with lectures in the mornings, the afternoons being free for discussion sessions.

It was felt that, to make the material covered available to a wider audience, it would be worthwhile to publish the proceedings of these workshops as 'The ICTP Series in Theoretical Physics'. The current volume is the first such, and is based on the 1983 particle physics workshop on $N = 1$ Supergravity.

March 1984 Abdus Salam

ABSTRACT

A review of recent developments in the applications of N=1 Supergravity to the construction of unified models of elementary particle interactions is given. Couplings of N=1 Supergravity with matter consisting of an arbitrary set of left handed multiplets and a gauge multiplet are discussed. General formulation of spontaneous symmetry breaking and the criteria for breaking of internal symmetry and of local supersymmetry are described. Construction of specific Supergrvity GUT models, the gauge heirarchy in GUT models and a deduction of the low energy effective potential are discussed. The phenomena of SU(2)XU(1) electro-weak gauge invariance breakdown by Supergravity in tree models and by radiative corrections using renormalization group methods are described. Model independent formulations of low energy physics which encompass tree and the renormalization group methods of breaking SU(2)XU(1), but also allow more general scenarios are discussed. The particle content of Supergravity unified theories at low energy is given which includes predictions, for a class of supergravity models, of light gauge fermins, i.e. of a charged Dirac fermion, the Wino, with a mass below the W boson mass and a neutral fermion, the Zino, below the mass of the Z boson mass. "Direct" gaugino masses arising from loops for the photino and the gluino, (which are massless at the tree level) are exhibited. Decays of the W and Z into photino, Wino, and Zino modes and their branching ratios into various channels are given. Experimental signals for the supersymmetric decays are discussed and some prominent signatures such as decays into one and two jets with unbalanced energy and momentum are pointed out. The current status of the ρ-parameter, including supergravity GUT effects is given. Other experimental consequences of supergravity unified theories are also discussed.

Notation: Secs. VII and VIII use the Lorentz metric diag $\eta_{\mu\mu} = (-1, +1, +1, +1)$ and standard left handed Weyl spinors (projected by $P_- = (1 - \gamma_5)/2, \gamma_5^\dagger = \gamma_5$). The discussion of the supergravity - matter couplings, Sec. II and App. A, are in the notation of [C4] and [C7].

C O N T E N T S

I. INTRODUCTION

The intrinsic appeal of supersymmetry [A1-A3] is that it unifies bosons and fermions by placing them in common irreducible multiplets. It also possesses the remarkable feature that in unified theories with supersymmetry with intrinsically different mass scales, one can maintain a mass heirarchy without the necessity of fine tuning in each order of perturbation due to the well known "no renormalization" theorem [A4]. These features of supersymmetry have led to a great deal of interest for some time in the application of supersymmetry for model building of unified theories [B1-B27]. In spite of many of their attractive features, these SUSY models based on global supersymmetry do not always lead to phenomenologically acceptable low energy predictions due to the difficulty in breaking global supersymmetry. For example in certain SUSY GUT models [B4] there must exist at least one scalar boson with a mass less than a quark mass. In other SUSY theories of the low energy domain, one also finds unacceptable particle spectrum [B17]. In SUSY theories it is not possible to cancel the cosmological constant by adjustment of parameters intrinsic to the theory (i.e. without the addition of adhoc terms to the Lagrangian). Further in SUSY GUT theories [B14,B22] the GUT mass M is quite large i.e. $M \sim 10^{16}$ GeV so that $kM \sim 10^{-2}$ where

$$k = (8\pi G)^{1/2} = 0.41 \times 10^{-18} \mathrm{GeV} \quad (1.1)$$

and G is the Newtonian constant. This means that in SUSY GUT theories it is not so reasonable to neglect gravitational effects due to the close proximity of the GUT mass M and the Planck mass $M_p = k^{-1}$.

These considerations led the authors to propose a new class of models [D1-D3] based on N=1 Supergravity [C1-C4]. In these lectures we shall discuss the construction of these models and subsequent developments. As is obvious gravity is included into the grand unification in an intrinsic way. Contrary to the popular belief that the gravitational effects should be negligible at low energy here one finds [D1] that in supergravity unified theory, supergravitational phenomena proportional to k (square root of the Newtonian constant) act as the trigger for the breakdown of the SU(2)XU(1) Weinberg-Salam electro-weak symmetry [F1] and the scale of the electro-weak breaking characterized by m_W can be related to the gravitino mass m_g [D1] i.e.

$$M_W \sim 0\ (m_g)\ . \qquad (1.2)$$

Since m_g is proportional to k one finds that the supergravitional interactions far from being negligible appear to run physics in the low energy domain up to 10^2-10^3 GeV. One has then essentially a dynamical unification of the supergravitational and electro-weak phenomena. The origin of this dynamical unification is the spontaneous breaking of SU(2)XU(1) [D1] through the super-Higgs effect [C5-C7] very much like the dynamical unification of the electromagnetic and the weak phenomena in the Weinberg-Salam theory is achieved through spontaneous breaking by the ordinary Higgs effect.

Criteria for stability of the SU(2)XU(1) breaking scale have been given at the one loop level [D3,D7,D8] and explicit models which are one loop stable have been exhibited [D19,D20]. Further, the breaking of SU(2)XU(1) by the super-Higgs need not necessarily arise at the tree level ("tree breaking (T.B) models"). The SU(2)XU(1) breaking can also arise dynamically from renormalization group loop corrections [D5,D6,D16-D18] ("renormalization group (R.G.) models"). The essential difference structurally between the T.B. models and the R.G. models is the following: In the T.B. models, the low energy effective potential [D4,D10,D11], in the Higgs' sector contains a pair of Higgs doublets H^{α} and H'_{α}, $\alpha = 1,2$ and a singlet field U. The singlet plays a crucial role in SU(2)XU(1) breaking in the T.B. models. For the case of the R.G. models one discards the singlet and retains only the pair of Higgs doublets. SU(2)XU(1) is broken in the R.G. models due to the large Yukawa couplings of a heavy top quark. Typically one needs $m_t \gtrsim 100-200$ GeV. More recently model independent analyses of the low energy domain have been given which can accomodate the T.B. and the R.G. models as well as a whole class of other variations in between [E7-E9].

At the phenomenological level, Supergravity unified theories make some unique predictions and are free from the usual defects of the corresponding global SUSY theories. Thus for example, unlike the global supersymmetry theory [B4] there are no light scalar bosons in these theories. Indeed in the low energy domain the scalar bosons (that do not become superheavy) acquire a characteristic mass $O(m_g)$ [E10,E1]. Unlike the case of the SUSY theories, in supergravity theories, one can at least eliminate the cosmological constant by adjustment of a parameter (fine tuning) in the superpotential (see Sec. III for details). In supergravity unified theories there is a suppression of the flavor changing neutral currents due to a cancellation of m_g^2 in $(mass)^2$ differences of the bosonic partners of the quarks. These cancellations are not arranged by hand as in global theories but arise naturally as a consequence of the supergravity models.

The fermionic partners of the photon and the gluons, i.e. the photino and the gluinos, are massless at the tree level. However, the photino and the gluinos grow masses at the loop level [E6,D17].

$$\tilde{m}_\gamma = \frac{8}{3}\frac{\alpha}{4\pi}\bar{C}\,m_g \;;\; \tilde{m}_g = \frac{\alpha_3}{4\pi}\bar{C}m_g \approx 8\tilde{m}_\gamma \tag{1.3}$$

where $\bar{C}$ is proportional to the Casimir of the (heavy) multiplet exchanged in the loop (See Eq. 5.13). For normal size heavy multiplets, one expects $\tilde{m}_\gamma$ to lie in the range of (1–10) GeV and the gluino mass in the range (5–80) GeV from Eq. (1.3).

The fermionic partners of the W and Z bosons also grow masses at the tree level. In fact as discussed by Weinberg [E4] and the authors [E5] there appear in theories of the type discussed above, relatively light gauge fermions. Thus in the charged sector one has a Wino (Supersymmetric partner of the W boson), the $\tilde{W}_{(-)}$, lying below the W boson and in the neutral sector one has a Zino (Supersymmetric partner of the Z boson), the $\tilde{Z}_{(-)}$, lying below the Z boson. In each of these sectors there also exist additional supersymmetric partners, i.e. a Wino, $\tilde{W}_{(+)}$, lying above the W boson and a Zino, $Z_{(+)}$, lying above the Z boson, in Supergravity models [E5] as well as other neutral Zinos [E7–E9].

The existence of $\tilde{W}_{(-)}$ and $\tilde{Z}_{(-)}$ are clearly very exciting from an experimental viewpoint since one has the possiblity of detecting these particles in the decays of the W and Z which have now been experimentally confirmed [F2,F3]. The decays of the W and Z which are universal are for the W [E4]

$$W^{\pm} \rightarrow \tilde{W}_{(-)}^{\pm} + \tilde{\gamma} \tag{1.4}$$

and for the Z [E5]

$$Z \rightarrow \tilde{W}_{(-)}^{+} + \tilde{W}_{(-)}^{-} \tag{1.5}$$

since they occur in both the T.B. and the R.G. models. A remarkable feature of the decay of Eq. (1.5) is that it is of "Industrial Strength" i.e. the branching ratio of $Z \rightarrow \tilde{W}^{+} + \tilde{W}^{-}$ relative to $Z \rightarrow e^{+}e^{-}$ is characteristically of size O(5). The T.B. and the R.G. models each possess additional decays which are unique to them. Thus in the T.B. model one has the decay [E3,E9]

$$W^{\pm} \rightarrow \tilde{W}_{(-)}^{\pm} + \tilde{Z}_{(-)} \tag{1.6}$$

if $M_W > (\tilde{m}_W + \tilde{m}_Z)$. In R.G. models, the process of Eq. (1.6) is kinematically disallowed. However, in the R.G. models there exists a new light neutral Higgsino $\tilde{Z}_{(3)}$ (we call it the "Twilight Zino" due to its very weak coupling with ordinary matter) which couples with normal strength with the Z allowing for the following decay [E8,E9]

$$Z \rightarrow \tilde{Z}_{(3)} + \tilde{Z}_{(3)} \; . \tag{1.7}$$

Each of the decays of Eqs. (1.4)-(1.7) have their own characteristic signal. Thus the decay of Eq. (1.4) would lead to jets in one direction balanced by an

Unidentified Fermionic Object (UFO), the photino, in the opposite direction. These UFO events would each consist of a single jet with unbalanced momentum. There are similar characteristic UFO signals for other processes of Eqs. (1.5)-(1.7). Further, there exists the possibility of testing the T.B. vs the R.G. models through Eqs. (1.6) and (1.7) in the decays of the W and Z. The W and Z decays thus provide a possible test of supersymmetry at the $\bar{p}p$ collider. A search for gluinos can also be carried out at the $\bar{p}p$ collider [E11,E5]. Other possible tests of supersymmetry could come through e^+e^- collisions in the processes $e^+e^- \to \tilde{\gamma}\tilde{\gamma}\gamma$ [E13,E14]. (See Fig. (1)) This experiment can be carried out at the current energies at PEP and PETRA. In most models the production of the selectron (the supersymmetric partner of the electron) would require larger energies such as those contemplated at LEP (though the model of [E14] can accomodate a light selectron). The process $e^+e^- \to \tilde{\gamma}\tilde{\gamma}e^+e^-$ [E15] offers another possibility, though its cross-section is expected to be quite small (see Fig. (2)).

There are cosmological constraints on some of the "ino" masses [E16,E17] though in these lectures we shall not discuss these here in any detail.

II. SUPERGRAVITY MATTER COUPLINGS AND EFFECTIVE POTENTIAL

In our analysis we shall use N=1 Supergrvity with the minimal set of auxiliary fields [C1-C4]. Here the field content consists of the spin2, spin3/2 fields $e_{a\mu}$, ψ_μ and the auxiliary fields S,P,Aμ. The Supergravity Lagrangian invariant (up to a total divergence) under local supersymmetry transformations is then [C1,C2]

$$L_{S.G.} = -\frac{e}{2k^2}R(e,\omega) - \frac{e}{3}\,|u|^2 + \frac{e}{3}A_\mu A^\mu - \frac{1}{2}\,\bar{\psi}_\mu R^\mu \qquad (2.1)$$

where

$$u = S - iP \qquad (2.2)$$

$$R_{\mu\nu}{}^{rs} = \partial_\mu\omega_\nu{}^{rs} + \omega_\mu{}^{rp}\omega_{\nu p}{}^{s} - \mu \leftrightarrow \nu \qquad (2.3)$$

$$R^\mu = \varepsilon^{\mu\nu\rho\sigma}\gamma_5\gamma_\nu D_\rho(\omega)\psi_\sigma \qquad (2.4)$$

$$R = e_r{}^\mu\; e_s{}^\nu R_{\mu\nu}{}^{rs} \qquad (2.5)$$

$$D_\mu = \partial_\mu + (1/2)\;\omega_{\mu rs}\sigma^{rs} \qquad (2.6)$$

$$\omega_{\mu rs} = \omega_{\mu rs}\;(e) + K_{\mu rs}\;(e,\psi_\mu) \qquad (2.7)$$

$$K_{\mu rs}(e,\psi_\mu) = (k^2/4)(\bar{\psi}_\mu\gamma_r\psi_s - \bar{\psi}_\mu\gamma_s\psi_r + \bar{\psi}_r\gamma_\mu\psi_s) \qquad (2.8)$$

and e is the determinant of the vierbein.

In the construction of the GUT models one needs couplings of N=1 Supergravity with matter. Matter consists of left-handed chiral (F-type) multiplets

$$\Sigma^a = (Z^a, \chi^a{}_L, h^a) \tag{2.9}$$

where $Z^a = A^a + iB^a$ are complex scalar fields, $\chi^a{}_L$ are left-handed Weyl spinors and h^a are complex auxiliary fields. The index a in Σ^a is an internal symmetry index such that Σ^a belongs to a reducible representation of a gauge group G. In addition, matter contains a vector (D-type) multiplet $V(V=V^\dagger)$ which has components

$$V = (C, \xi, H, K, V_\mu, \lambda, D) \tag{2.10}$$

and is in the adjoint representation of the gauge group G. In (2.9) ξ, λ are Majorana spinors, C,H,K are scalars while D is an auxiliary scalar field. The vector multiplet is reduced significantly in the Wess-Zumino guage [A1] and one has

$$V = (V_\mu, \lambda, D) \ . \tag{2.11}$$

The rules of coupling a single F-type multiplet and a single D-type multiplet with Supergrvity are known. For the F-type multiplet one has [C2,C3]

$$e^{-1}L_F = \mathrm{Re}[h + uZ + \bar{\varphi}_\mu \gamma^\mu \chi + \bar{\varphi}_\mu \sigma^{\mu\nu} \varphi_{\nu R} Z] \tag{2.12}$$

and for the D-type multiplet one has [C3]

$$\begin{aligned} e^{-1}L_D = {} & D - \frac{ik}{2}\bar{\varphi}_\mu \gamma^5 \gamma^\mu \lambda - \frac{2}{3}(SK - PH) \\ & + \frac{2}{3}kV_\mu (A^\mu + \frac{3}{8}ie^{-1}\varepsilon^{\mu\rho\sigma\tau}\bar{\varphi}_\rho \gamma_\tau \varphi_\sigma) \\ & + i\frac{k}{3}e^{-1}\bar{\xi}\gamma_5 \gamma_\mu R^\mu + \frac{ik^2}{8}\varepsilon^{\mu\nu\rho\sigma}\bar{\varphi}_\mu \gamma_\nu \varphi_\rho \bar{\xi}\varphi_\sigma \\ & - \frac{2}{3}k^2 C\, e^{-1}L_{S.G.} \end{aligned} \tag{2.13}$$

One may contrast the results on Eqs. (2.12) and (2.13) with the corresponding sitution for global supersymmetry where only the F and D terms are admissible in the Lagrangian. For the case of supergravity all elements of the F and D multiplets enter the Lagrangian to preserve the local supersymmetry gauge invariance.

Cremmer et. al [C7] have given the most general coupling of a single chiral multiplet with supergravity. This scheme exhibits the possibility of spontaneous breakdown of the Supergravity gauge invariance and a mass growth for the gravitino through a minimization of its true effective potential. However, the existence of only a single chiral multiplet in the coupling scheme does not allow one to formulate supergravity GUT theories. This led the authors to a generalization of these results to couple N=1 Supergravity to an arbitrary number of chiral multiplets belonging to a reducible representation of a grand unified gauge group G and simultaneously to a gauge multiplet belonging to the adjoint representation of the gauge group [C8]. Equivalent formulations have been given by other authors [C9–C13]. The details of construction are presented in Appendix A and in this section we shall only outline the general procedure and state results for the relevant parts of the Lagrangian we need.

The procedure consists in forming the most general F and D multiplets out of Eqs. (2.9) and (2.11) using the rules of tensor calculus and maintaining the invariance under the gauge group G. Next one couples these F and D multiplets to supergravity using Eqs. (2.12) and (2.13). The resulting Lagrangian is expressed most conveniently in terms of functions $g(Z^a)$, $\phi(Z^a, Z_a)$ and $f_{\alpha\beta}(z^a) \cdot g(z)$ is the familiar superpotential which is the lowest element of the most general gauge singlet F-multiplet formed out of the chiral multiplets of Eq. (2.9) i.e.

$$g(z^a) = \sum B_{a_1 \ldots a_m} Z^{a_1} \ldots Z^{a_m} \tag{2.14}$$

$\phi(Z^a, Z_a)$ represents the lowest element of the most general gauge singlet D multiplet formed out of the chiral multiplet Σ^a of Eq. (2.9) and its hermitian conjugate, i.e.,

$$\phi(Z^a, Z_a) = \sum A^{a_1 \ldots a_m}{}_{b_1 \ldots b_n} Z_{a_1} \ldots Z_{a_m} Z^{b_1} \ldots Z^{b_n} \; . \tag{2.15}$$

The co-efficients $B_{a_1 \ldots a_m}$ and $A^{a_1 \ldots}{}_{b_n}$ in Eqs. (2.14) and (2.15) are arbitrary parameters except that they are chosen to maintain invariance under G. A convenient procedure in implmenting the coupling scheme is to first couple the gauge multiplet V of Eq. (2.11) to the chiral multiplets Σ^a and next couple the resultant structure to supergravity.

To obtain the Lagrangian in a useful form one must first carry out an elimination of all the auxiliary fields in the theory (both in the matter and the supergrvity sector). In addition one needs to make point transformations to put the kinetic energy of the dynamical fields into canonical form. The details of the resulting Lagrangian are given in Appendix A.

The Lagrangian is determined in terms of two arbitrary functions $\mathcal{G}(Z^a,Z_a)$ and $f_{\alpha\beta}(Z^a)$ The function $\mathcal{G}(Z^a,Z_a)$ is a special combination of ϕ and g:

$$\mathcal{G}(Z^a,Z_a) = -\frac{k^2}{2}\, d(Z,Z^\dagger) - \ln(\frac{k^6}{4}|g(Z^a)|^2) \qquad (2.16)$$

and exhibits the invariance

$$g \rightarrow g\, e^{f(z)} \qquad (2.17)$$

$$d \rightarrow d - \frac{2}{k^2}(f(z) + f^\dagger(z^\dagger)) \qquad (2.18)$$

where

$$d \equiv -(6/k^2)\ln(-\frac{k^2}{3}\phi) \quad . \qquad (2.19)$$

Eqs. (2.17) and (2.18) along with the fact that the kinectic energies of the scalar fields are proportional to $\mathcal{G},^a{}_b \equiv \partial^2\mathcal{G}/\partial Z^b\partial Z_a$ makes $\mathcal{G},^a{}_b$ as the metric in the Kähler manifold [C14] defined by coordinates Z^a,Z_b.

The arbitrary function $f_{\alpha\beta}(Z)$ enters in the Yang-Mills sector [C9]

$$e^{-1}L(F_{\alpha\beta}) = \frac{1}{2}f_{\alpha\beta}\Big(-\frac{1}{4}F^{\alpha}_{\mu\nu}F^{\mu\nu\beta} - \frac{1}{2}\bar{\lambda}^{\alpha}D\!\!\!/\lambda^{\beta} + \frac{1}{2}D^{\alpha}D^{\beta}$$

$$+ \frac{1}{4}F^{\alpha}_{\mu\nu}\tilde{F}^{\beta}_{\mu\nu} - \frac{1}{2}D_{\mu}(\bar{\lambda}^{\alpha}\gamma^{\mu}\lambda_{R}^{\beta})\Big) + \text{h.c.} \tag{2.20}$$

There is no theory to determine $f_{\alpha\beta}(z)$ in the current framework. However, we know that in the absence of gravitational interactions renormalizability requires that one have $f_{\alpha\beta} = \delta_{\alpha\beta}$. Thus if the quantum supergravity theory was appropriately controlled in the ultra-violet domain such as through the phenomena of "asymptotic safety" [F4], the deviations of $f_{\alpha\beta}$ from the global limit should be only due to gravitational loop effects. Weinberg [E4] has argued that these loops obey to a good approximation, a U(n) symmetry among the n chiral multiplets, and as a consequence, deviations of $f_{\alpha\beta}$ from $\delta_{\alpha\beta}$ should be very small. We shall thus assume in our analysis

$$f_{\alpha\beta}(Z) = \delta_{\alpha\beta} \quad . \tag{2.21}$$

Under the assumption of Eq. (2.21) the Bose part of the Lagrangian then takes the form

$$L_B = -(e/2k^2)R(e,\omega) + (e/k^2)\mathcal{G}_{,}{}^{a}{}_{b}\,\mathcal{D}_{\mu}Z_{a}\mathcal{D}^{\mu}Z^{b}$$

$$- (e/k^4)\exp(-\mathcal{G})[3 + (\mathcal{G}^{-1})^{a}{}_{b}\,\mathcal{G}_{,a}\,\mathcal{G}_{,}{}^{b}]$$

$$-\frac{1}{4}e\,F^{\alpha}_{\mu\nu}F^{\mu\nu\alpha}-\frac{e}{8k^4}|g_\alpha \mathcal{G},_a(T^\alpha Z)^a|^2 \quad . \tag{2.22}$$

It is useful to express Eq. (2.22) in an alternate form using the function $d(Z,Z^\dagger)$ of Eq. (2.19). One has then L_B in the form

$$e^{-1}L_B=-\frac{e}{2k^2}R(\omega(e))-\frac{1}{2}d,^a{}_b\mathcal{D}_\mu Z_a\mathcal{D}_\mu Z^b-\frac{1}{4}F^\alpha_{\mu\nu}F^{\mu\nu\alpha}-e^{-1}V \tag{2.23}$$

where V is the potential of the scalar fields and is given by

$$V=\frac{e}{2}\exp\left(\frac{k^2}{2}d\right)[(d^{-1})^a{}_bG_aG^b-\frac{3}{2}k^2|g|^2] \tag{2.24}$$

where G_a is defined by

$$G_a\equiv\frac{\partial g}{\partial Z^a}+\frac{k^2}{2}d,_a g \tag{2.25}$$

and $(d^{-1})^a{}_b$ is the inverse of the matrix $(d),^b{}_a\equiv\partial^2d/\partial Z^a\partial Z_b$.

In our new notation, $d(Z,Z^\dagger)$ acts as the potential in the Kähler manifold. The choice

$$d=Z_aZ^a \tag{2.26}$$

corresponds to a flat Kähler manifold with the matrix $d,^b{}_a=\delta^b{}_a$ and leads to a normalized kinetic energy for the scalar fields in Eq. (2.23). The choice of Eq. (2.24) may be too restrictive and the gravitational loop corrections may possibly modify Eq. (2.24). However, the gravitational loop corrections to a good approximation are still expected to preserve the U(n) symmetry among the n-chiral fields and a more general choice for an effective Kähler potential d should be a general function of $Z_a Z^a$. For simplicity, however, we shall carry out most of the analysis in the following sections under the assumption of a flat Kähler manifold, Eq. (2.26).

III. SPONTANEOUS SYMMETRY BREAKING AND SUPER-HIGGS EFFECT

For the flat Kähler manifold the extrema equations arising from Eq. (2.24) have the form

$$\left(\frac{\partial G_b}{\partial Z_a} + \frac{k^2}{2} Z^a G_b\right) G^b - k^2 g\, G = 0 \ . \tag{3.1}$$

On the real manifold of VEVS, Eq. (3.1) reduces down to

$$T_{ab}\, G_b = 0 \tag{3.2a}$$

where G_b is given by Eq. (2.25) and T_{ab} is defined by

$$T_{ab} = \frac{\partial^2 g}{\partial Z^a\, \partial Z^b} + \frac{k^2}{2}\left(Z_a \frac{\partial g}{\partial Z^b} + Z_b \frac{\partial g}{\partial Z^a}\right) + \frac{k^4}{4} Z_a Z_b g - k^2 \delta_{ab} g \ . \tag{3.2b}$$

Eq. (3.2) may be satisfied through the vanishing of all the G_a i.e.

$$G_a = 0 \ . \tag{3.3}$$

Under this circumstance one has that supersymmetry is unbroken though the gauge symmetry of the theory may be broken. An example of such a breaking is provided by the superpotential

$$g = \lambda\left(\frac{1}{3}\mathrm{Tr}\Sigma^3 + \frac{1}{2}M\,\mathrm{Tr}\Sigma^2\right) \ . \tag{3.4}$$

Where $\Sigma_y{}^x$ is the adjoint representation 24 of SU(5). Satisfaction of Eq. (3.3) shows that a minimum of the potential exists when $(\Sigma_y{}^x)_{diag}$ possesses one of the following vacuum expectation values

$$\text{(i) } 0 \quad \text{(ii) } \tfrac{1}{3}M(1,1,1,1,-4) \quad \text{(iii) } M(2,2,2,-3,-3) \tag{3.5a}$$

and the VEVS of all other components of Σ_y^x are zero. Solution in Eq. (3.5a) does not break supersymmetry. However, solutions (ii) and (iii) of Eq. (3.5a) break the gauge group. Thus solution (ii) breaks SU(5) into SU(4)XU(1) while solution (iii) breaks SU(5) into SU(3)XSU(2)XU(1). A second example is provided by the superpotential [B2]

$$g = \lambda_o X(M^2 - \mathrm{Tr}\Sigma^2) + \lambda_1 \mathrm{Tr}\Sigma^2\Delta + \lambda'_1 Y \mathrm{Tr}\Sigma\Delta \quad . \tag{3.5b}$$

Here Δ_y^x is a second $\underline{24}$ representation of SU(5). Satisfaction of Eq. (3.3) shows that a minimum exists when $X = \Delta_y^x = 0$ and Σ_y^x and Y have one of the following two solutions

$$\text{(i)} \quad \Sigma_y^x = \frac{M}{\sqrt{20}}(\delta_y^x - 5\,\delta_5^x\,\delta_y^5),\ Y = \frac{3M}{\sqrt{20}} \tag{3.6}$$

$$\text{(ii)} \quad \Sigma_y^x = \frac{M}{\sqrt{30}}[2\delta_y^x - 5(\delta_4^x\delta_y^4 + \delta_5^x\delta_y^5)],\ Y = \frac{M}{\sqrt{30}}\frac{\lambda_1}{\lambda'_1} \quad . \tag{3.7}$$

Again solutions of Eqs. (3.6) and (3.7) preserve supersymmetry but the gauge symmetry is broken. Solution (ii) corresponds to a residual symmetry of SU(4)XU(1) while (iii) corresponds to the residual symmetry SU(3)XSU(2)XU(1).

The symmetry breaking solutions of the type Eq. (3.5) which preserve supersymmetry exhibit an interesting phenomena. Substitution of Eq. (3.3) in the tree effective potential gives

$$V_{min}(Z_o{}^a, Z_{oa}) = -\frac{3k^2}{4}|g(Z_o{}^a)|^2 \exp(\frac{1}{2}k^2 Z_{oa}\, Z_o{}^a) \ . \qquad (3.8)$$

Eq. (3.8) implies that the degeneracy of the vacuum solutions encountered in global supersymmetry is removed due to the $O(k^2)$ corrections to the vacuum energy [D1,D22,D23]. From Eq. (3.8) one finds that if any one of the minimum solutions is chosen to be Minkowskian by the adjustment of an additive constant to the superpotential, then all other solutions would necessarily be of anti-deSitter nature and would have vacuum energies which are negative. Normally a situation where the Minkowskian vacuum arises in association with anti-deSitter vacuum would seem to require that Minkowskian vacuum would be unstable. However, the presence of gravitation can help restore stability [D24]. In fact for situations where supersymmetry is preserved, Weinberg [D22] has argued that the Minkowskian vacuum would actually be stable for any finite size perturbation even though it does not have the lowest energy.

One may notice that for the potential of Eq. (3.5b), both solutions (3.6) and (3.7) give a vanishing $g(Z_o{}^a)$ which implies that for this case gravitation does not lift the degeneracy of models of the type Eq. (3.5b) [D2]. Thus there exists the possibility in these models of realizing a vacuum structure where the Minkowskian vacuum is the lowest state of energy when supersymmetry is broken [D2].

We consider next the case of broken supersymmetry. On the mass-shell supersymmetry transformations for the spin 3/2 and spin 1/2 Weyl fields are (see Appendix A)

$$\delta\chi^a = (\gamma^\mu \varepsilon_R)\hat{D}_\mu Z^a - \frac{1}{2}(\mathcal{G},_b\bar{\varepsilon}\chi^b - \mathcal{G},^b\bar{\chi}_b\varepsilon)\chi^a + (\mathcal{G}^{-1})^a{}_b\,\mathcal{G},^b{}_{cd}\,\bar{\chi}^c\chi^d\varepsilon_L$$

$$- k^{-1}\exp(-\frac{\mathcal{G}}{2})(\mathcal{G}^{-1})^a{}_b\mathcal{G},^b\varepsilon_L \qquad (3.9)$$

$$\delta\psi_{\mu L} = 2k^{-1}D_\mu(e,\psi_\mu)\varepsilon_L + k^{-2}\exp(-\frac{\mathcal{G}}{2})\gamma_\mu\varepsilon_R + \frac{1}{2}(\mathcal{G},_a\bar{\varepsilon}\chi^a - \mathcal{G},^a\bar{\chi}^a\varepsilon)\psi_{\mu L}$$

$$+ k^{-1}\sigma_{\mu\nu}\varepsilon_L\mathcal{G},^a{}_b\,\bar{\chi}^a\gamma^\nu\chi^b - \frac{k}{8}(\delta^\nu_\mu + \gamma^\nu\gamma_\mu)\varepsilon_L\bar{\lambda}^\alpha\gamma_\nu\gamma_5\lambda^\alpha - \frac{k^{-1}}{2}(\mathcal{G},_a D_\mu Z^a - \mathcal{G},^a D_\mu Z_a)\varepsilon_L \qquad (3.10)$$

From Eqs. (3.9) and (3.10) one finds for the vacuum expectation values (for a flat Kähler manifold)

$$\langle\delta\chi^a\rangle_o = -\exp(\frac{k^2}{4}Z_o{}^aZ_{oa})G_a(Z_o)\varepsilon_L \qquad (3.11)$$

$$\langle\delta\psi_{\mu L}\rangle_o = \frac{k}{2}\exp(\frac{k^2}{4}Z_o^aZ_{oa})g(Z_o)\gamma_\mu\varepsilon_R + 2k^{-1}\partial_\mu\varepsilon_L\ . \qquad (3.12)$$

Thus a necessary requirement for the breaking of supersymmetry is that at least one G_a is non-zero. In the representation where T_{ab} is diagonal this implies at least one non-vanishing eigenvalue for T_{ab}. In the unitary gauge where the spin1/2 Goldstino is absorbed by the gravitino, the gravitino mass is given by [C7]

$$m_g = \frac{k^2}{2}g(Z_o)\exp(\frac{k^2}{4}Z_o^\dagger Z_o)\ . \qquad (3.13)$$

The simplest example of the super-Higgs effect occurs when one couples a single chiral multiplet $\Sigma = (Z, \chi_L, h)$ with Supergravity and the superpotential is of the form [C6]

$$g_2(Z) = m^2(Z + B) \tag{3.14}$$

where m^2 and B are constants. Here $G(Z) \neq 0$, and hence Eq. (3.2) requires $T_{zz} = 0$. One finds [C6,C7]

$$kZ_{(-1)} = a(\sqrt{2} - \sqrt{6});\ kB_{(-)} = -a(2\sqrt{2} - \sqrt{6});\ a = \pm 1 \tag{3.15}$$

where the condition on $B_{(-1)}$ is chosen so that $V_{min} = 0$. From Eq. (3.15) we note that the super-Higgs field Z has VEV $\sim O(M_p)$. Eq. (3.14) represents the simplest possibility and one may consider the more general case of an arbitrary super-Higgs potential of the form

$$g_2(Z) = m^2k^{-1}f_2(kZ) \tag{3.16}$$

where $f_2(kZ)$ is an arbitrary dimensionless function of kZ with the expansion

$$f_2(kZ) = f_2{}^{o} + kZf_2{}^{1} + \ldots \tag{3.17}$$

The $T_{zz} = 0$ condition for superpotentials of type in Eq. (3.16) would yield supersymmetry breaking solutions which are characteristically of the form

$$\langle Z\rangle \sim 0(k^{-1});\ \langle g_2\rangle \sim 0(k^{-1}m^2)\ . \tag{3.18}$$

Eq. (3.18) together with Eq. (3.13) then implies

$$m_g = \frac{k^2}{2}g_2(\langle Z\rangle)\exp(\frac{k^2}{4}\langle Z^\dagger Z\rangle) \sim 0(m^2k)\ . \tag{3.19}$$

In Sec. V we shall identify the gravitino mass as characteristically of the size of the weak interaction scale i.e. $m_g \sim 0(M_W)$ [D1]. Using this correspondence we can identify the mass scale m entering Eq. (3.16) of the super-Higgs potential. One finds from Eq. (3.19) that

$$m \sim (M_W\ M_{Planck})^{1/2} \sim 10^{10}GeV\ . \tag{3.20}$$

This means that the mass scale entering the super-Higgs effect is an intermediate mass scale and is related by a geometric heirarchy to the weak interaction mass scale and the Planck mass.

We note in passing that there are additional constraints which a super-Higgs potential must satisfy in order that it be an admissible potential for realistic model building. For example, an important requirement is that the particle spectrum of the theory after spontaneous breakdown contain no tacheonic modes. A quadratic form for the superpotential $g_2(z) = m(z^2 + B)$ can be excluded on this basis.

IV. SUPERGRAVITY MODELS

In setting up realistic supergravity models we shall find it convenient to classify the full set of fields Z^A in the matter sector into two categories: the field Z in the super-Higgs sector and the remaining matter fields Z^a. Thus we write

$$Z^A = (Z^a, Z) \ . \tag{4.1}$$

Our basic supergravity model is then defined by the superpotential [D1]

$$g(Z^A) = g_1(Z^a) + g_2(Z) \ . \tag{4.2}$$

In the limit k=0, the dynamics of the fields Z^a and the field Z are completely disjoint. However, for k non-zero, the two sectors interact through supergravitational interactions, and the dynamics of each sector is affected. The most dramatic effect occurs in the sector of the fields Z^a due to the influence of the field Z in that one finds the appearance of soft-breaking in the Z^a-sector due to the super-Higgs effect [D1,D4]. We shall illustrate the soft-breaking phenomena by the very simple example where

$$g_1(Z^a) = 0 \ . \tag{4.3}$$

For the case of Eq. (4.3) one finds that the potential

$$V = \frac{1}{2}\exp(\frac{k^2}{2}Z_A\ Z^A)[G_A G^A\ - \frac{3}{2}k|g|^2] + \frac{e}{32}|g_\alpha(Z_a, (T^\alpha Z)_a]^2 \tag{4.4}$$

gives the mass term $\frac{1}{2}m_g^2\ Z_a\ Z^a$ to scalar fields Z^a while for the corresponding fermionic mass matrix (see Appendix A)

$$m_{ab} = \exp(\frac{k^2}{4}Z_a\ Z^a)[\frac{\partial^2 g}{\partial Z^a \partial Z^b} + \frac{k^2}{2}(Z_a\ \frac{\partial g}{\partial_Z b} + Z_b\ \frac{\partial g}{\partial_Z a})$$

$$+ \frac{k^4}{12}Z_a Z_b g - \frac{2}{3}\frac{\partial g}{\partial Z^a}\ \frac{\partial g}{\partial Z^b}g^{-1}]+\text{h.c.} \tag{4.5}$$

one finds for Eq. (4.3) the result $m_{ab} = 0$. Thus the degeneracy between the bosons and the fermions is lifted and the mass of the gravitino characterises the scale by which the degeneracy is broken. Indeed one notices that the super-Higgs effect generates a common mass which is equal to the gravitino mass in this approximation.

In the general analysis one has $g_1 \neq 0$ which makes the general analysis for the computation of soft breaking more complex. One of the reasons for this complexity is that in general some of the fields in Z^a may be super-heavy involving the GUT mass scale M which is close to the Planck mass k^{-1}. For a SUSY theory $M \sim 3\text{x}10^{16}$ GeV and so one has

$$\varepsilon \equiv kM \sim 10^{-2} \ . \tag{4.6}$$

This means that in the solution to the minimization equations for the determination of VEVS, higher order k corrections of size

$$(kM)M,\ (kM)^2M, \ldots (kM)^6M \tag{4.7}$$

must be controlled in order that the low energy theory be protected from the GUT mass scale M. This is a new heirarchy problem arising only in supergravity models and has no direct analogue in the corresponding global theories. To account for these new complexities of the light and the heavy fields in the matter sector Z^a we classify our fields as follows:

$$Z_a = (\{Z_i\},\{Z_i'\},\{Z_\alpha\}) \ . \tag{4.8}$$

The Z_i are fields with VEVS of O(M) and also of masses of O(M). The Z_i' have vanishing or small $O(m_g)$ VEVs but have masses of O(M). The fields Z_α are light fields with VEVs and mass of $O(m_g)$. Our purpose next is to examine the minimization conditions and establish criteria which would generate the above gauge heirarchy at the tree level. Generation of such a pattern of gauge heirarchy is motivated by our desire for developing GUT models where one needs gauge heirarchies of type Eq. (4.8).

In order to examine the full heirarchy problem at the tree level we develop an expansion solution for the VEVs in powers of k:

$$Z_a = Z_a^{(0)} + Z_a^{(1)} + Z_a^{(2)} + \ldots \tag{4.9}$$

$$Z_{i'}^{(0)} = 0, \ Z_\alpha^{(0)} = 0 \tag{4.10}$$

$$Z = Z^{(-1)} + Z^{(0)} + \ldots \tag{4.11}$$

where we have that $Z_A^{(n)} \sim O(k^n)$. It is useful to rescale the fields,

$$z_i = M^{-1}Z_i,\ z_\alpha = m_s{}^{-1}Z_\alpha,\ z = kZ \tag{4.12a}$$

where

$$m_s \equiv km^2 \tag{4.12b}$$

so that z_i, z_α, z etc. are dimensionless and have expansions similar to Eqs. (4.9) – (4.11) but beginning at zeroth order in k. It is also useful to define the dimensionless quantities $\bar{g}$, $\bar{G}_a$ and $\bar{G}_z$ as follows:

$$\bar{g} = (k/m^2)g,\ \bar{G}_z = m^{-2}G_z,\ \bar{G}_a = m_s{}^{-2}G_a \tag{4.13}$$

Now from the extrema equations that determine the VEVs, one can show [D10] that the desired tree gauge heirarchy would be destroyed if $\bar{G}_a$ contained terms of size M/m_s or k^{-1}/m. Indeed one can establish that for a wide class of theories which obey the restriction

$$g_{,\alpha i} \sim O(m_s) \tag{4.14}$$

$$g_{,\alpha\beta} \sim O(m_s) \tag{4.15}$$

one has $\bar{G}_i$, $\bar{G}_\alpha$ and $\bar{G}_z$ are of order unity so that generally one has at the minimum of the effective potential

$$G_a \sim 0(m_s^{\ 2}) \tag{4.16}$$

with corrections to the leading order term which are very small i.e. $\varepsilon\delta_s$ and $\delta_s^{\ 2}$ where ε is defined by Eq. (4.6) and

$$\delta_s \equiv km_s \sim 10^{-16} \ . \tag{4.17}$$

Normally one would expect $G_i \sim M^2$ on dimensional grounds and so the result of Eq. (4.16) is quite remarkable. It is Eq. (4.16) which plays the central role in guaranteeing protection of VEVS in the tree level minimization equations. The essential meaning of Eqs. (4.16) is that the effects of the GUT sector characterized by the GUT mass M and of the super Higgs sector characterized by the Planck mass on the low mass sectors is only of size $0(m_s)$ which maintains the mass heirarchy. Thus typically in the low mass sectors the full solution of the extrema equations taking account of the GUT and super Higgs sector would generte the following type series expansion for the light field VEVS:

$$Z_\alpha = m_s z_\alpha = m_s z_\alpha^{\ (0)} + A_\alpha m_s (kM)(km_s) + B_\alpha m_s (km_s)^2 + \ldots \tag{4.18}$$

where $z_\alpha^{\ (0)}, A_\alpha, B_\alpha, \ldots$ are dimensionless numbers of order unity. This tree level protection holds to arbitrary orders in k.

Eqs. (4.14) and (4.15) act as essential constraints necessary to achieve the tree level gauge heirarchy in the construction of realistic supergravity GUT models. Thus certain types of couplings must either be eliminated or unnaturally suppressed in the superpotential. Thus for example the coupling

$\lambda Z_i Z_j Z_\alpha$ can only appear in the superpotential provided $\lambda \sim m_s/M$ while the coupling $\lambda' Z_i Z_j Z$ can appear provided $\lambda' \sim k m_s$.

For model building it is found useful to eliminate the super-Higgs and the heavy fields to achieve a low energy effective potential [D4,D10,D11]. Thus consider the effective potential of the full theory $V(Z_i; Z_\alpha, Z)$ which obeys the extrema equations

$$\frac{\partial V}{\partial Z} = 0; \; \frac{\partial V}{\partial Z_i} = 0 \quad . \tag{4.19}$$

One may solve Eqs. (4.19) to express Z and Z_i in terms of Z_α i.e.

$$Z = Z[Z_\alpha]; \; Z_i = Z_i[Z_\alpha] \quad . \tag{4.20}$$

In practice Eqs. (4.20) would be exhibited in a power series in k. We note that in the low energy domain we are for the present only interested in recovering operators of dimensionality four or less to construct the low energy effective potential [which turns out to imply that the series expansion of Eq. (4.20) need not go beyond order (m_s) corrections]. Insertion of Eq. (4.20) into Eq. (4.19) then gives the low energy effective potential $U(Z_\alpha)$

$$U(Z_\alpha) = V(Z_i[Z_\alpha]; \; Z_\alpha; \; Z[Z_\alpha]) \, . \tag{4.21}$$

The protection conditons Eqs. (4.14) and (4.15) guarantee that (Z_α) has terms only $O(m_g{}^4)$ and terms $O(M m_g{}^3)$, $O(M^2 m_g{}^2)$ etc. cancel [D11]. An alternate procedure is to eliminate the heavy fields and super Higgs fields in the extrema equations of the light sector. Since in the extrema equations, protection of the low energy mass scale has already been achieved, the integration of these equations would yield an effective potential which has the low energy protection already built in. The relevant equations to integrate are Eqs. (3.1) in the light sector. One finds [D10]

$$U(Z_\alpha, Z^\alpha) = \frac{1}{2}\exp(\frac{k^2}{2}|Z_o|^2)[\tilde{g}^\dagger_{1,\alpha}\tilde{g}_{1,\alpha} + m_1^2 Z_\alpha Z^\alpha + (\omega + \omega^\dagger) + m_s^2 (\tilde{g}_{1,i}G_i^{(0)} + h.c.)] \tag{4.22}$$

where

$$\omega = m_2\tilde{g}_1 + m_3 Z^\alpha \tilde{g}_{1,\alpha} \tag{4.23}$$

and

$$\tilde{g}_1(Z_i, Z_\alpha) = g_1(Z_i, Z_\alpha) - g_1(Z_i, 0) - b \ . \tag{4.24}$$

In the deduction of Eq. (4.22) we have used the form of Eq. (4.2) for the superpotential. The three mass parameters m_1, m_2 and m_3 reduce to two when the cosmological constant condition is imposed since then $m_1 = |m_3|$. The constants are

$$m_1^2 = \frac{1}{2}m_s^2[\bar{G}_z^{(0)}\bar{G}_z^{(0)*} - \bar{g}_2^{(0)}\bar{g}_2^{(0)*}] \tag{4.25}$$

$$m_2 = \frac{1}{2}m_s[z^{(0)}\bar{G}_z^{(0)} - 3\bar{g}_2^{(0)}] \tag{4.26}$$

$$m_3 = \frac{1}{2}m_s\bar{g}_2(0),\ m_1 = |m_3| \ . \tag{4.27}$$

In Eq. (4.24), the Z_i appearing are evaluated using the extrema equations for the heavy sectors to zeroth and first order:

$$Z_i = Z_i^{(0)} - \frac{1}{2}m_s\bar{g}_2^{(0)}(M^{-1})_{ij}Z_j^{(0)} \equiv Z_i^{(0)} + Z_i^{(1)} \ . \tag{4.28}$$

When one imposes the vanishing of the cosmological constant condition, Eq. (4.22) reduces to the following

$$U(Z_\alpha, Z_\alpha^\dagger) = \frac{1}{2}\exp(\frac{k^2}{2}|Z_o|^2)[g_{1,\alpha}\tilde{g}^\dagger_{1,\alpha} + m_1^2 Z_\alpha Z^\alpha$$

$$+ (m_2 g'_1 + m_3 Z_\alpha g_{1,\alpha} + \text{h.c.})] \tag{4.29}$$

where

$$g_1' = g_1(Z_i^{(0)}, Z^\alpha) \ .$$

The results of Eq. (4.29) are equivalent to the analysis of [D11]. The analysis of [D11] is carried out for a general Kähler manifold obeying the U(n) symmetry and involves two additional mass parameters. The analysis of [D4] is limited only to the elimination of the super-Higgs fields and the heavy fields are not integrated out.

V. SU(2)XU(1) BREAKING BY SUPERGRAVITY

A remarkable aspect of supergravity models is that one may induce the breakdown of SU(2)XU(1) gauge invariance through supergravitational interactions [D1] and there exist now many models which contain realization of such a breakdown [see Sec. D of References]. We shall illustrate this aspect of supergravity unified theories in a tree model. We choose for our superpotential g_1 the following form

$$g_1 = \lambda_1(\frac{1}{3}\mathrm{Tr}\Sigma^3 + \frac{M}{2}\mathrm{Tr}\Sigma^2)$$
$$+ \lambda_2 H(\Sigma_x{}^y + 3M'\delta_x{}^y)H'_y + \lambda_3 UH'_x H^x$$
$$+ \varepsilon_{uvwxy} H^u M^{vw} f_1 M^{xy} + H'_x M^{xy} f_2 M'_y + B_1 \ . \qquad (5.1)$$

Here $\Sigma_x{}^y, H^x$, H'_x are left handed chiral fields in the 24, 5 and $\bar{5}$ representations. M'_x and M^{xy} are the matter (quark-lepton) $\bar{5}$ and 10 superfields and f_1 and f_2 are Yukawa coupling constant matrices in the generation space. The superpotential of Eq. (5.1) has the structure of Eq. (3.4) in the GUT sector. In Sec. III we found that in the scheme of Eq. (3.4), one had three inequivalent minima after spontaneous breaking with residual symmetries of (i) SU(5), (ii) SU(4)xU(1) and (iii) SU(3)XSU(2)XU(1) corresponding to the three solutions of Eq. (3.5). Of course the physically interesting vacua are those corresponding to the case (iii) in Eq. (3.5) and this is the solution we choose for our analysis. Further to guarantee that the doublets of Higgs are light we must impose the condition M'=M. On using

Eq. (4.22) one then has the effective potential in the low energy domain in the form [D10]

$$U = \frac{1}{2}E_o[m_3{}^2|U|^2 + m_3{}^2(1+9\lambda^2)(\bar{H}_\alpha H^\alpha + H'_\alpha \bar{H}'^\alpha)$$

$$- 6\lambda m_3{}^2(H'_\alpha H^\alpha + \bar{H}_\alpha \bar{H}'^\alpha) - a\sqrt{\frac{3}{2}}m_3 z^{(0)}\lambda_3 \cdot$$

$$(UH'_\alpha H^\alpha + U^*\bar{H}_\alpha\bar{H}'^\alpha) - 3m_3\lambda_3\lambda(U+U^*)(\bar{H}_\alpha H^\alpha + H'_\alpha\bar{H}'^\alpha) + (\lambda_3)^2|U|^2(\bar{H}_\alpha H^\alpha + H'_\alpha\bar{H}'^\alpha)$$

$$+ (\lambda_3)^2(H'_\alpha H^\alpha)\bar{H}_\beta\bar{H}'^\beta)] + U_M; \qquad \bar{H}^\alpha \equiv H^{\alpha\dagger} \text{ etc.} \tag{5.2}$$

where $E_o = \exp(4-2\sqrt{3})$. In Eq. (5.2), U_M is the part which involves the Yukawa interactions of the squark and the selepton fields. In the following analysis we shall examine only the minima arising from the Higgs part of the potential in Eq. (5.2) so that the VEVS of the squarks and the selepton fields are assumed to be zero. In the analysis of the extrema equations that govern the VEVs of the Higgs fields appearing in Eq. (5.2), it is convenient to introduce dimensionless parameters x and y defined by

$$U = -axm_s/(\sqrt{2\lambda 3}); \; H^\alpha = \delta_5{}^\alpha y m_s/(\sqrt{2\lambda 3}) \; . \tag{5.3}$$

One finds for the extrema equations then the following

$$\xi_1 y^2 + 2xy^2 + x = 0 \tag{5.4}$$

$$y[x^2 + \xi_1 x + y^2 + \xi_2{}^2] = 0 \tag{5.5}$$

where

$$\lambda \equiv \lambda_2/\lambda_1; \ \xi_1 = 3-\sqrt{3 - 6\lambda}; \ \xi_2 = 1-3\lambda \ . \tag{5.6}$$

An examination of Eqs. (5.4) – (5.5) show that solutions exist on two branches as shown below:

$$\text{(i)} \quad \lambda \gtrsim 1.05: \ SU(5) \rightarrow SU(3)XSU(2)XU(1) \rightarrow SU(3)^C XU_\gamma(1) \tag{5.7}$$

$$\text{(ii)} \ \lambda \lesssim \frac{2}{3}(1.05): \ SU(5) \rightarrow SU(3)XSU(2)XU(1)$$
$$\rightarrow SU(2)^C XSU(2)XU(1) \ . \tag{5.8}$$

On branch (2), one finds that for a range of values of the parameter λ, SU(2)XU(1) can break spontaneously to $U_\gamma(1)$. On this branch $SU(3)_C$ is preserved to all orders in k. On branch (ii), for a range of values of λ, SU(2)XU(1) is exactly preserved but $SU(3)_C$ is broken. The physically interesting branch is, of course, given by (i). The scale of breakdown of SU(2)XU(1) is given by m_s of Eq. (4.12b). Thus m_s must be the size of electro-weak mass scale i.e. $O(m_W)$. From Eq. (5.3) one finds on using the experimental value for the Higgs VEV, the result

$$m_s = \xi(246)\text{GeV}; \ \xi \equiv \frac{\lambda_3}{y} \sim O(1) \ . \tag{5.9}$$

Additional theoretical input is needed to determine ξ which would help define the gravitino mass uniquely. Further, the inclusion of squark and selepton fields into the analysis may generate new minima which may lie lower [D21]. However, from the viewpoint of constructing realistic models what is required is not that our Minkowskian vacuum be the lowest in energy but rather that it be stable against decay into the lower minima or at least its decay life be much larger than the observed life of the universe. Further in certain models of SU(2)XU(1) breaking it is possible to construct vacua where the pathologies discussed above may be circumvented. Thus in Ref. [D2] T.B. models were exhibited where the Minkowskian vacuum may also be made the lowest state of energy and in Ref. [D17] R.G. models are constructed where the inclusion of non-vanishing VEVs of the squark and selectron fields do not generate new minima which lie lower.

The breakdown of SU(2)XU(1) discussed above is truly induced by supergravitational interactions. Thus the VEVs of the Higgs fields in Eq. (5.3) are non-zero for case (i) which breaks SU(2)XU(1) due to supergravitational interactions proportional to k. Thus as $k \to 0$, m_s vanishes and the SU(2)XU(1) symmetry is restored. We note in passing that the model of SU(2)XU(1) breaking proposed in [D4] does not satisfy this criteria i.e. the breakdown of SU(2)XU(1) does not require supergravitational interactions since even as $k \to 0$ one has a breakdown of SU(2)XU(1) in the underlying global theory.

The model of Eq. (5.1) possesses the gauge heirarchy at the tree level to all orders in k. At the one loop level additional constraints are needed to guarantee the gauge heirarchy [D3,D7,D26]. As first noted in [D7] the loop gauge heirarchy is destroyed in the model of Eq. (5.1) due to the coupling structure $\lambda_3 UH'H$ [D27]. However, one loop stability criteria may be satisfied by the introduction of additional multiplets as recently discussed in [D20,D19].

Next we turn to the structure of supergravity GUTS. The full analysis of the particle spectrum of supergravity GUTS shall be discussed later in a model independent framework. Here the only mass spectra we shall discuss are those of the photino and the gluinos.

(a) "Direct" Gaugino Masses

The photino and the gluinos are massless at the tree level. It has been suggested [D25] that the gravitational radiative corrections may generate massess for the gluinos due to a term in the supergravity-matter Lagrangian of the type

$$\frac{k^2}{8}\bar{\lambda}\gamma_\mu\sigma^{\nu\rho}\varphi^\mu\bar{\varphi}_\nu\gamma_\rho\lambda \ . \tag{5.10}$$

Such a term would generate a one loop mass to the gauginos of order $O(m_g)\Lambda^2/M_p{}^2$ where Λ is the ultra-violet cut-off i.e. $\Lambda = M_p$ (see Fig. (3a)). However, there exists another part of the Lagrangian which is of the form

$$\frac{k}{4}\bar{\lambda}\gamma_\mu\sigma^{\nu\rho}\varphi^\mu F_{\nu\rho} \tag{5.11}$$

and its loop contribution (see Fig. (3b)) cancels the leading $O(mg)\Lambda^2/M_p^{\,2}$ term arising from Eq. (5.10). Thus the gravitational loop corrections do not appear to generate significant loop gaugino masses. (A similar conclusion appears in Ref. [E18]). Actually a source of significant "direct" gaugino masses is due to the exchange of heavy fields of the GUT sector (see Fig. (4)). The relevant interactions thus involve the couplings of the gauginos with GUT chiral multiplets. The basic interaction is

$$L_{int} = ig_\alpha[\bar{\chi}^a(\frac{T^\alpha}{2})^a{}_b Z^b\lambda^\alpha - \bar{\lambda}^\alpha Z_a(\frac{T^\alpha}{2})^a{}_b\chi^b] \quad . \tag{5.12}$$

The gaugino mass matrix (for the exchange of real representations) is determined by

$$m_\alpha = \frac{g_\alpha^{\,2}}{16\pi^2}m_g\,\bar{C} \tag{5.13}$$

where $\bar{C}$ = CD(R)/D(A), C is the Casimir, D(R) the dimensionality of the representation exchanged and D(A) is the dimensionality of the adjoint representation. (When the exchanged representation is also adjoint e.g. $\Delta_y^{\,x}$ is a 24 of SU(5) one has $\bar{C}$ = C). Eq. (1.3) follows from Eq. (5.13) [E6,D19,D17]. The exchange of quark and lepton multiplets do not generate any significant contributions to the gaugino masses. Their combined contribuitions are typically less than a GeV [E9,E18].

Photino masses enter importantly in cosmological considertions. The mass density of the universe has an upper limit of of approximately $2\text{X}10^{-29}$ g/cm^3.

A lower bound on stable heavy-neutral lepton masses arises because the cosmic-density arising from these particles cannot exceed the current mass density of the universe [E19]. It has recently been pointed out [E17] that Majorana fermion annihilation rate is P-wave suppressed. This effect tends to increase the lower bound on the cosmologically allow photino masses compared to the conventional lower bounds of 1-2GeV. The lower bound of $\tilde{m}_\gamma$ of 7 GeV was found in [E17]. Similar cosmological considerations can also be carried out for the Twilight Zino showing that it cannot be the lowest lying odd R-parity fermion.

(b) The ρ Parameter

In the electroweak theory the parameter ρ is defined as the ratio of the neutral current to the charge current Fermi couplings. In the SU(2)XU(1) theory with doublets of Higgs ρ=1 at the tree level. However, deviations from unity arise due to the electroweak loop corrections. A significant source of contribution to the ρ-parameter was pointed out by Veltman [E20] as arising from the mass-splittings of the third generation (t,b) quark doublet.

$$\Delta\rho = \left(\frac{3}{\sqrt{2}}\frac{G_F}{8\pi^2}\right)m_t^2 \tag{5.14}$$

which appears to set an upper bound on the top quark mass of ~O(400) GeV due to the experimental bound on ρ [E21]

$$\rho = 1.01 \pm .02 \text{ [Experiment]} . \tag{5.15}$$

It is thus interesting to investigate what the status of the ρ parameter is in supergravity GUT theories. A full analysis of ρ for T.B. Supergravity theories has recently been carried out by Eliasson [E22]. (The ρ parameter for a supergravity theory but neglecting vertex corrections is also discussed in [E23]). The results of [E22] are shown in Fig. (5). The contributions of the top quark to ρ are positive. The vertex corrections to ρ are found to be generally small. The upper limit on the top quark mass consistent with Eq. (5.15) is O(250) GeV [E22]. Of course for a large top quark mass, the renormalization group effects due to its Yukawa couplings can no longer be neglected and one must carry out a more extensive analysis to determine the true upper bound.

VI. RADIATIVE BREAKING OF SU(2)XU(1)

In the preceeding Supergravity GUT models, SU(2)XU(1) breaking occurred at the tree level, and required the presence of a singlet field U. We consider in this section an alternate class of models in which no singlet field is present (so that SU(2)XU(1) remains unbroken at the tree level) but radiative corrections [calculated using the renormalization group (R.G.)] trigger the breaking of SU(2)XU(1). In order for radiative corrections to be sufficiently strong, it generally is necessary to postulate the existance of a heavy top quark (i.e. $m_t \gtrsim 100$ GeV for light direct (loop) gaugino masses, and $m_t \gtrsim 55$ GeV even with very heavy direct masses.) These models [D5,D6,D16–D18] are similar to earlier global SUSY models [B8–B11,B24,B25], but differ from many of the latter which stress the need for large gaugino masses to act as a supersymmetry breaking agent. In contrast Supergrvity GUT models break supersymmetry by the super Higgs effect, a phenomena which imposes a unique set of boundary conditions at the GUT mass, without imposing any a priori restrictions on the gaugino masses.

We consider here the simplest model containing one pair of Higgs doublets H^a and H'_a, $a = 1,2$ (H^a interacts with the t-quark in the Yukawa terms) and will retain only the t-quark Yukawa interactions in the following (since only those are large). The total low energy effective potential consists of a Higgs and t-quark part, $U = U_H + U_t$ and takes the form [suppressing the exponential factor of Eq. (4.22)]:

$$U_H = [m_H{}^2\bar{H}^\alpha H^\alpha + m_{H'}{}^2\bar{H}'_\alpha H'_\alpha + \bar{m}_3 m_g(H'_\alpha H^\alpha + \bar{H}'_\alpha\bar{H}^\alpha)]$$

$$+ \frac{1}{8}g_1{}^2(\bar{H}^\alpha H^\alpha - \bar{H}'_\alpha H'_\alpha)^2 + \frac{1}{8}g_2{}^2(\bar{H}\tau_i H - H'\tau_i\bar{H}')^2 \quad (6.1)$$

$$U_t = h_t{}^2|\tilde{\bar{t}}_R\tilde{t}_L|^2 + h_t{}^2|\tilde{\bar{t}}_R H|^2 + m_g A_t h_t(\tilde{\bar{t}}_R H^2\tilde{t}_L + h.c.)$$

$$+ m_{\tilde{t}_R}{}^2\tilde{\bar{t}}_R\tilde{t}_R + m_{\tilde{t}_L}{}^2\tilde{\bar{t}}_L\tilde{t}_L \quad (6.2)$$

where $\bar{H}^\alpha \equiv H^{\alpha\dagger}$ etc, $\tilde{t}_L$ and $\tilde{t}_R$ are the top squark chiral multiplet partners of the l.h. Weyl top doublet and r.h. top singlet fields, and g_i, i=1,2,3 and the SU(3), SU(2), U(1) gauge coupling constants. In the tree approximation, $m_H{}^2 = m_{H'}{}^2 = m_{\tilde{t}_R}{}^2 = m_{\tilde{t}_L}{}^2 = m_g{}^2$ (where m_g is of the size of the gravitino mass) and are the universal scalar masses while $\bar{m}_3 \sim O(m_g)$. (For an arbitrary Kähler manifold, m_3 is not simply related to m_g.) The $H'_\alpha H^\alpha$ coupling term arises from either translations of the heavy VEVs of the GUT sector, or possibly is an effective low energy structure from loop corrections involving the GUT sector. We assume here that $\bar{m}_3$ is considerably smaller than m_H or $m_{H'}$ [D16–D18]. The trilinear coupling in Eq. (6.2) arises from the elimination of the super Higgs field [D1,D4,D11,D10]. The coupling constant A_t is determined by the choice of super Higgs potential in the tree approximation (e.g. $A_t = 3 - \sqrt{3}$ for the Polony case).

The R.G. method proceeds as follows. Eqs. (6.1) and (6.2) are assumed to take on their tree values at the GUT mass M. At lower energy scales μ, the masses and coupling constants become running coupling constants, determined by the renormalization group equations. Thus SU(2)XU(1) breaking will occur if at some scale μ_o the top quark couplings turn m_H^2 negative. In the limit of small m_3, the minimum then occurs at

$$v \equiv \langle H^2 \rangle \cong 2|m_{H'}|(g_1^2 + g_2^2)^{-1/2}, \quad w \equiv \langle H'_2 \rangle \approx -\bar{m}_3 m_g v/(m_H^2 + m_{H'}^2) \qquad (6.3)$$

and hence

$$m_H^2(\mu_o) = -\frac{1}{2}M_z^2 . \qquad (6.4)$$

Note that generally $w \ll v$ for these models since m_3 is small.

The renormalization group equations for supersymmetric theories have been worked out by a number of authors [B24,D50,D17]. Neglecting the direct gaugino masses, the leading terms read

$$\frac{d}{dt}\Psi = (\alpha_t/2\pi)M\Psi + (A_t^2\alpha_t/2\pi)m_g\Psi_1; \quad t \equiv \ln(\frac{\mu}{M}); \quad \alpha_t = h_t^2/4\pi \qquad (6.5)$$

where the collumn symbol $\Phi = (m_H^2, m_{\tilde{t}_R}^2, m_{\tilde{t}_L}^2)$, and the β function matrix M and Φ, are given by

$$M = \begin{pmatrix} 3 & 3 & 3 \\ 2 & 2 & 2 \\ 1 & 1 & 1 \end{pmatrix} \quad ; \Psi_1 = \begin{pmatrix} 3 \\ 2 \\ 1 \end{pmatrix} \qquad (6.6)$$

$\alpha_t(\mu\ )$ and $A_t(\mu)$ obey

$$\frac{d}{dt}\alpha_t = (3\alpha_t^2/\pi) - (\alpha_t/\pi)(\frac{8}{3}\alpha_3 + \frac{3}{2}\alpha_2 + \frac{13}{18}\alpha_1) \qquad (6.7)$$

$$\frac{d}{dt}A_t = (3\alpha_t/\pi)A_t . \qquad (6.8)$$

In order to get a rough idea of what these equations imply we first neglect the gauge coupling terms of Eq. (6.7). Then Eqs. (6.5)-(6.7) can be solved analytically. Thus Eqs. (6.7) and (6.8) yield

$$\alpha_t = \frac{\alpha_o}{1-\xi}\ ;\ A_t = \frac{A_o}{1-\xi}\ ;\ \xi(t) = 3\alpha_o t/\pi \tag{6.9}$$

where α_o, A_o are the top coupling constants at the GUT mass M. (A_o is determined by the choice of super Higgs potential e.g. $A_o=3-\sqrt{3}$ for the Polony model). To solve Eq. (6.5), it is convenient to expand Ψ in terms of the eigenvectors of M i.e. $\Psi = c_1\Psi_1 + C_2\Psi_2 + C_3\Psi_3$ where

$$M\Psi_1 = 6\Psi_1;\ M\Psi_2 = 0 = M\Psi_3;\ \Psi_2 = (1,-2,1),\ \Psi_3 = (1,0,-1)\ . \tag{6.10}$$

Since at the GUT masses the tree boundary conditons hold, m_H^2, $m_{\tilde{t}_R}^2$ and $m_{\tilde{t}_L}^2$ all equal m_g^2 there and hence one finds

$$C_1(\xi) = \frac{1}{2}\frac{m_g^2}{1-\xi}[1 + \frac{1}{3}A_o^2\frac{\xi}{1-\xi}];\ C_2(\xi) = 0;\ C_3(\xi) = -\frac{1}{2}m_g^2 \tag{6.11}$$

which yields

$$m_H^2 = 3C_1 - \frac{1}{2}m_g^2;\ m_{\tilde{t}_R}^2 = 2C_1;\ m_{\tilde{t}_L}^2 = C_1 + \frac{1}{2}m_g^2 \tag{6.12}$$

we see that for the physical range of ξ ($-\infty < \xi \lesssim 0$) if $C_1(\xi) > 0$ (i.e. A_o^2

is not too large) The squark masses $m_{\tilde{t}_R}^{\ 2}$, $m_{\tilde{t}_L}^{\ 2}$ can never turn negative. However, the Higgs mass clearly can (for ξ sufficiently negative) signalling the breaking of SU(2)XU(1). Using Eqs. (6.4), (6.11), (6.12) and the fact that at the minimum of the effective potential $h_t = m_t/v$, one finds the condition on the top quark mass m_t to be:

$$\frac{m_t^{\ 2}}{v^2} = \frac{4\pi^2}{3} \frac{1}{(-t_o)} [1 - \frac{2B}{\{(A_o^{\ 2} + 2B - 3)^2 + 4B(3-B)\}^{\frac{1}{2}} + 3 - A_o^{\ 2}}] \tag{6.13}$$

where v = 177 GeV, $B = 1 - M_z^{\ 2}/m_g^{\ 2}$ and $t_o = \ln(\mu_o(M))$. (One has $\mu_o = M_W$ and $M \sim 3\text{x}10^{16}$ GeV.)

It is interesting to trace the origin of the above spontaneous breaking. From Eqs. (6.6) and (6.12) one sees that it is the mixing of the Higgs mass with the squark masses in the β-function matrix M combined with the boundary conditions at the GUT mass that allows $m_H^{\ 2}$ to turn negative. <u>These boundary conditions are unique to Supergravity GUT theories</u> and have no known analogue in global SUSY theories. (The boundary conditions relate the gravitino mass to SU(2)XU(1) phenomena. The additional soft breaking term proportional to $A_t^{\ 2}$ (also unique to Supergravity GUTS aids the SU(2)XU(1) breaking, but is not the dominant effect). [In fact, if A_o is too large $\tilde{m}_{tR}^{\ 2}$ turns negative destabilizing the physical vacuum as can be seen from Eqs. (6.11), (6.12)]. We also note from Eq. (6.9), A_t is reduced at the low energy regime from its GUT value A_o, though not dramatically so, and so t-squark soft breaking terms may have interesting physical consequences at low energies.

For the Polony choice $A_o = 3 - \sqrt{3}$, Eq. (6.13) requires that 80GeV $\lesssim m_t \lesssim$ 115 GeV. The gauge couplings of Eq. (6.7) tend to inhibit the spontaneous breaking and if these are included one finds [D17,D18] (for general A_o) that 100 GeV $\lesssim m_t \lesssim$ 195 GeV. Finally, if one includes the direct gaugino mass terms $\tilde{m}_a$ they tend to aid the breaking of SU(2)XU(1) and one has the lower bound [D16–D18] $m_t \gtrsim$ 55GeV (in the limit $\tilde{m}_a \to \infty$).

VII. MASS SPECTRUM-MODEL INDEPENDENT ANALYSIS

The fact that supersymmetry breaks at a relatively low mass in Supergravity GUTs [i.e. $m_g \sim (M_W)$] suggests the existance of low lying supersymmetric particles accessible to experiment. This possibility represents one of the most exciting aspects of the theory. As discussed by Weinberg and the authors [E1,E4,E5]. There apper in most models relatively light gauge fermions (supersymmetric partners of the SU(3)XSU(2)XU(1) gauge bosons) lying below the W and Z bosons. In the limit where the direct (loop) gaugino masses may be neglected, Weinberg [E4] has shown that if the U(n) symmetry of gravitational loops is valid there will always be at least one charged Wino, $\tilde{W}^+$ (partner of the W boson) lying <u>below</u> the W boson, and one neutral Zino, $\tilde{Z}^o$ (partner of the Z boson) lying below the Z boson as well as a light photino $\tilde{\gamma}$. These may therefore become detectible in such decays as

$$W^{\pm} \rightarrow \tilde{W}^{\pm} + \tilde{\gamma}$$

$$W^{\pm} \rightarrow \tilde{W}^{\pm} + \tilde{Z}$$

$$Z^o \rightarrow \tilde{W}^+ + \tilde{W}^- \qquad (7.1)$$

In this section we analyse the mass spectrum of the gaugino and other sectors of the theory and do this in a <u>model independent</u> fashion, [E7, E9, E12] i.e. in a general way that encompasses a wide class of interesting models.

All models currently considered assume the existance of a pair of Higgs doublet superfields $\hat{H}^{\alpha}$ and $\hat{H}'_{\alpha}$, $\alpha = 1,2$. In addition the tree breaking models assume the presence of a singlet superfield $\hat{U}$. In the low energy domain, after integrating out the heavy fields and eliminating the super Higgs field, these fields must interact in a renormalizable way [D11, D10] and thus can be characterized simply by an effective low energy superpotential $g_{eff} = g + g_M$ where

$$g = \mu\, \hat{H}'_{\alpha}\hat{H}^{\alpha} + \lambda'\hat{U}\hat{H}'_{\alpha}\hat{H}^{\alpha} - \frac{1}{6}\lambda''\hat{U}^3 \tag{7.2}$$

and g_M contains the Yukawa interactions of the matter multiplets and the Higgs doublets.

In minimizing the low energy effective potential discussed in Sec. IV, both H'_2 and H^2 may develop VEVs (as well as U in the T.B. models). We thus parameterize this breaking by a single angle α [E7–E9] defined by

$$\tan\alpha \equiv w/v;\ v = \langle H^2\rangle,\ w = \langle H'_2\rangle \quad . \tag{7.3}$$

Hence $M_W = \frac{1}{2}g_2(v^2 + w^2)^{1/2}$ and $M_z = M_W/\cos\theta_W$. Our general theory thus depends on the parameters α, μ, λ' and λ'' (as well as m_g) and different models can be characterized by different domains of these parameters. Thus for the tree breaking models of Sec. V one has that α is close to 45° and the other parameters are large i.e.

$$\text{(T.B.)} \qquad \alpha \approx 40^{\circ}\text{–}50^{\circ};\ \mu \sim m_g;\ \lambda',\ \lambda'' \sim 1 \tag{7.4}$$

while for the renormalization group models of Sec. VI, $\alpha \sim \mu$ (m_g is small and λ' and λ'' do not eneter i.e.

$$\text{(R.G.)} \qquad \alpha \sim 10^{\circ}\text{–}25^{\circ};\ \mu/m_g \ll 1;\ \lambda' = 0 = \lambda'' \tag{7.5}$$

(though recently an R.G. model has been proposed [D26] with $\alpha \cong 45^{\circ}$, $\mu \sim m_g$ and $\lambda' = 0 = \lambda''$). Thus the formalism is broad enough to deal with all cases.

We consider first the fermion mass matrices. In the low energy sector, the fermion fields are (a) the SU(3)XSU(2)XU(1) Majorana gauginos [$\lambda_r(x)$ (r=1...8), $\lambda^i(x)$ (i=1,2,3), and $\lambda^0(x)$], (b) the l.h. Weyl Higgsinos [$\tilde{H}^{\alpha}(x)$,$\tilde{H}'_{\alpha}(x)$, α=1,2] and (c) the neutral Weyl spinor of the $\hat{U}$ multiplet [$\tilde{U}(x)$]. Fermi mass terms arise from three possible sources: (i) From the superpotential [see Eq. 4.24)]:

$$L = -\frac{E}{2}(\overline{\chi^{ac}} g,_{ab}\chi^b + \text{h.c.}),\ E \equiv \exp(\tfrac{1}{2}\kappa^2\langle z\rangle^2) \tag{7.6}$$

where χ^a are the Weyl spinor components of the chiral multiplets, (ii) From the gaugino gauge interaction:

$$L_{\lambda\chi} = -\bar{\lambda}^{\alpha} m_{\alpha a}\chi^a + \text{h.c.} \tag{7.7}$$

where

$$m_{\alpha a} = i g_\alpha Z_b \left(\frac{T^\alpha}{2}\right)^b{}_a \qquad (7.8)$$

T^α are the group generators and Z_a are the scalar chiral partners of the χ^a, and (iii) the direct gaugino masses of Eq. (5.13):

$$L_\lambda = -\frac{1}{2}\bar{\lambda}^\alpha \tilde{m}_\alpha \lambda^\alpha \ . \qquad (7.9)$$

(a) <u>Charged Gaugino-Higgsino Fermion States</u>

The charged fermion fields, $\lambda \equiv (\lambda' - i\,\lambda^2)/\sqrt{2}$ and the charged Higgsinos $\tilde{H}^1$, $\tilde{H}'_1$ can conveniently be re-expressed in terms of two Dirac fields

$$\Psi_1 = \lambda_R + i\tilde{H}^1; \; \Psi_2 = \lambda_L + i\tilde{H}'_1{}^c \qquad (7.10)$$

where $\lambda_{R,L}$ are the r.h., l.h. components of λ. In the two component space labeled by $\Psi_o = (\Psi_1, \Psi_2)$, the charged mass matrix is

$$M = \mu_+ + \mu_- \tau_3 + \frac{1}{2}(\mu + \tilde{m}_2)\tau_1 + i\tau_2\gamma_5 \frac{1}{2}(\mu - \tilde{m}_2) \qquad (7.11)$$

where τ_a are Pauli matrices in Φ space,

$$\sqrt{2}\mu_\pm = M_W(\cos\alpha \pm \sin\alpha) \qquad (7.12)$$

and $\tilde{m}_2 = 3\tilde{m}\gamma/(8\sin^2\theta_W)$ where $\tilde{m}_\gamma$ is given by Eqs. (1.3) and (5.13).

One may easily diagonalize Eq. (7.11) by an "isotopic" and γ_5 transformation yielding the following mass eigenvalues and physical fields [E9,E7,E8]

$$\tilde{m}_{\pm} = \frac{1}{2}\left|[4\nu_+^2 + (\mu - \tilde{m}_2)^2]^{1/2} \pm [4\nu^2 + (\mu+\tilde{m}_2)^2]^{1/2}\right| \tag{7.13}$$

and

$$\tilde{W}_+ = i\cos\gamma_-\tilde{H}^1 + i\sin\gamma_+ \tilde{H}'{}_1{}^c - \sin\gamma_-\lambda_L + \cos\gamma_+\lambda_R$$

$$\tilde{W}_- = -i\sin\gamma_-\tilde{H}^1 - i\cos\gamma_+ \tilde{H}'{}_1{}^c - \cos\gamma_-\lambda_L + \sin\gamma_+\lambda_R \tag{7.14}$$

where

$$\tan2\beta_\pm = (\mu\mp\tilde{m}_2)/(2\nu_\pm);\ \gamma_\pm = \beta_+ \pm \beta_- \ . \tag{7.15}$$

The equation for $\tilde{W}_-$ holds for $\sin2\alpha \geq \mu\tilde{m}_2/M_W^2$. For $\sin2\alpha < \mu\tilde{m}_2/M_W^2$ replace $\tilde{W}_-$ by $\gamma_5\tilde{W}_-$.

Note that Eq. (7.13) implies

$$\tilde{m}_+\tilde{m}_- = |\sin2\alpha M_W^2 - \mu\tilde{m}_2| \tag{7.16}$$

and thus except when $\mu\tilde{m}_2$ is large, there is always one Wino, $\tilde{W}_-$ with mass $\tilde{m}_-$ $< M_W$. Such a particle may be considerably below the W.

(b) Neutral Gaugino-Higgsino Fermion States

In dealing with the neutral gaugino and Higgsinos, λ^3, λ^0, $\tilde{H}^2$, $\tilde{H}'_2$, $\tilde{U}$ it is convenient to introduce the following Majorana combinations

$$\lambda^\gamma = \sin\theta_W \lambda^3 + \cos\theta_W \lambda^0$$

$$\lambda^z = \sin\theta_W \lambda^0 - \cos\theta_W \lambda^3$$

$$\xi = i[\cos\alpha(\tilde{H}^2 - \tilde{H}^{2c}) - \sin\alpha(\tilde{H}'_2 - \tilde{H}'^{c}_2)]$$

$$\eta = -i[\sin\alpha(\tilde{H}^2 - \tilde{H}'^{c}_2) + \cos\alpha(\tilde{H}'_2 - \tilde{H}'^{c}_2)]$$

$$u = i(\tilde{U} - \tilde{U}^c) \ . \tag{7.17}$$

For this representation the direct gaugino masses are

$$\tilde{m}_z = \cos^2\theta_W \tilde{m}_2 + \sin^2\theta_W \tilde{m}_1 \cong 1.5\tilde{m}_\gamma$$

$$\tilde{m}_{\gamma z} = \cos\theta_W \sin\theta_W (\tilde{m}_1 - \tilde{m}_2) \cong -0.40\tilde{m}_\gamma \ . \tag{7.18}$$

The neutral mass matrix [E7-E9] is in general 5x5 for T.B. models and 4x4 for the R.G. models [which contains no singlet field $\hat{U}(x)$]. In the basis $\varphi = (\lambda^\gamma, \lambda^z, \xi, \eta, u)$ one has

$$M = \begin{pmatrix} \tilde{m}_\gamma & \tilde{m}_{\gamma z} & 0 & 0 & 0 \\ \tilde{m}_{\gamma z} & \tilde{m}_z & M_z & 0 & 0 \\ 0 & M_z & \mu\sin 2\alpha & \mu\cos 2\alpha & 0 \\ 0 & 0 & \mu\cos 2\alpha & -\mu\sin 2\alpha & \mu' \\ 0 & 0 & 0 & \mu' & \mu'' \end{pmatrix} \tag{7.19}$$

where [from Eq. (7.2)] $\mu' = \lambda' \ (v^2 + w^2)^{1/2}$ and $\mu'' = \lambda''\langle U\rangle$.

One may of course diagonalize Eq. (7.19) numerically. However, from Eq. (7.18) one sees that the gaugino mixing term $\tilde{m}_{\gamma z}$ is small, and neglecting this effect allows one to separate out the photino eigenfield λ^γ with eigenvalue $\tilde{m}_\gamma$. Further, in currently interesting models,

$$\mu^2 \cos^2 2\alpha \ll M_z{}^2 \tag{7.20}$$

(e.g. for T.B. models $\mu \sim M_z$ but $\alpha \approx 45^\circ$ while in R.G. models $\alpha \sim 15^\circ$ but $\mu \sim M_z/5$). In these approximations, Eq. (7.19) has the eigenvalues $\tilde{m}_\lambda$ and [E9]

$$\tilde{M}_\pm \cong [M_z{}^2 + \frac{1}{4}(\mu\sin 2\alpha - \tilde{m}_z)^2]^{1/2} \pm \frac{1}{2}(\mu\sin 2\alpha + \tilde{m}_z) \tag{7.21a}$$

$$\tilde{M}_{3,4} \cong [\frac{1}{4}(\mu\sin 2\alpha + \mu'')^2 + \mu'^2]^{1/2} \pm \frac{1}{2}(\mu\sin 2\alpha - \mu'') . \tag{7.21b}$$

(Detailed numerical analysis shows that the above approximations are quite good except in the R.G. model when there is an accidental degeneracy of light masses i.e. when $\mu\sin 2\alpha \cong -\tilde{m}_\gamma$.) The corresponding eigenfields (neglecting $\tilde{m}_{\gamma z}$ but not $\mu\cos 2\alpha$) are λ^γ and $\tilde{Z}_{(k)}$ $(k = +, -, 3,4,)$ where $\Psi_i = (\lambda^z, \xi, \eta, u)$ and

$$\tilde{Z}_{(k)} = (i\gamma_5)^{a_k} \sum \Psi_i N_{ik} D_k^{-1};\ a_+ = 0 = a_4,\ a_- = 1 = a_3 \tag{7.22}$$

Here

$$N_{1k} = M_z\,\tilde{\lambda}_k,\ N_{2k} = (\lambda_k - \tilde{m}_z)\tilde{\lambda}_k$$

$$N_{3k} = \mu\cos 2\alpha(\lambda_k - \tilde{m}_z);\ \tilde{\lambda}_k \equiv \lambda_k + \mu\sin 2\alpha - \mu'^2/(\lambda_k - \mu'')$$

$$N_{4k} = \mu\cos 2\alpha(\lambda_k - \tilde{m}_z)\mu'/(\lambda_k - \mu'')$$

$$D_k = [M_z^2\,\tilde{\lambda}_k^2 + (\lambda_k - \tilde{m}_z)^2\{\tilde{\lambda}_k^2 + \mu^2\cos^2 2\alpha(1 + \frac{\mu'^2}{(\lambda_k-\mu'')^2})\}$$

and

$$\lambda_\pm = \pm\tilde{\mu}_\pm,\ \lambda_3 = -\tilde{\mu}_3,\ \lambda_4 = \tilde{\mu}_4 \quad . \tag{7.23}$$

The eigenstates have different properties in the different models. In the T.B. model one sees from Eq. (7.21a) that again one generally has one Zino state, $\tilde{Z}_{(-)}$, lying below the Z boson i.e. $\tilde{\mu}_- < M_z$ and indeed can lie considerably below the Z. Note that this value of $\tilde{\mu}_-$ is indepdendent of the details of model i.e. of μ' and μ''. In general $\tilde{\mu}_{3,4}$ are relatively heavy. In contrast, in the R.G. models (where $\mu' = 0 = \mu''$) μ and α are small e.g. $\mu\sin 2\alpha \approx 10$GeV so both $\tilde{\mu}_\pm$ are relatively large and cluster very near (one above, one below) the Z. However, $\tilde{\mu}_3 \approx |\mu \sin 2\alpha|$ is quite light and well below the Z. We will refer to $\tilde{Z}_{(3)}$ as the "Twilight Zino", as it couples strongly to the Z boson but very weakly to all other matter

(c) Squark and Slepton States

Associated with each Weyl fermion is a complex scalar field e.g. for the first family of quarks and leptons, μ_L, μ_R, d_L, d_R, e_L, e_R, ν_L are the scalar fields $\tilde{u}_L$, $\tilde{u}_R$, $\tilde{d}_L$, $\tilde{d}_R$, $\tilde{e}_L$, $\tilde{e}_R$, $\tilde{\nu}_L$. Neglecting the small Yukawa interactins, these fields are in fact eigenstates of the squark and slepton mass matrices, and one finds for the mass eigenvalues [E9,E12]

$$m_{\tilde{\nu}}^2 = m_g^2 - \frac{1}{2}\cos 2\alpha M_z^2$$

$$m_{\tilde{e}_L}^2 = m_g^2 + (\frac{1}{2} - \sin^2\theta_w)\cos 2\alpha M_z^2$$

$$m_{\tilde{e}_R}^2 = m_g^2 + \sin^2\theta_w \cos 2\alpha M_z^2$$

$$m_{\tilde{u}_L}^2 = m_g^2 - (\frac{1}{2} - \frac{2}{3}\sin^2\theta\ )\cos 2\alpha M_z^2$$

$$m_{\tilde{d}_L}^2 = m_g^2 + (\frac{1}{2} - \frac{1}{3}\sin^2\theta_w)\cos 2\alpha M_z^2$$

$$m_{\tilde{u}_R}^2 = m_g^2 - \frac{2}{3}\sin^2\theta_w \cos 2\alpha M_z^2$$

$$m_{\tilde{d}_R}^2 = m_g^2 + \frac{1}{3}\sin^2\theta_w \cos 2\alpha M_z^2 \qquad (7.24)$$

which reduce to the results of [D17] for small α. In the R.G. models $\alpha \sim 10^\circ$–25°, and so the factor $\cos 2\alpha$ makes some correction.

In the T.B. models with $\alpha = 45^\circ$, the D term contributions proportional to M_z^2 vanish. If one includes the Yukawa interactions, they produce a 45° rotation in the squark and selectron states i.e. the eigenstates become $\tilde{u}_\pm = (\tilde{u}_L \pm \tilde{u}_R)/\sqrt{2}$ etc. The squark and selectron masses then become [E1,E5,D11,D19]

$$m(\tilde{q}_\pm)^2 = m^2_{\tilde{\gamma}} + m_q^2 \pm \beta m_{\tilde{\gamma}} m_q$$

$$m(\tilde{e}_\pm)^2 = m(\tilde{d}_\pm)^2;\ m^2_{\tilde{\gamma}} = m_g^2 \tag{7.25}$$

where m_q is the quark mass and β is a model dependent parameter of $O(1)$. Thus there is a small splitting of squark masses between generations produced by the Yukawa interactions proportional to the quark masses giving rise to a natural suppression of flavor changing neutral currents. All the squarks and selectrons are nearly degenerate with mass $\sim m_g = O(M_W)$.

(d) Higgs Bosons

All models contain one pair of Higgs doublets and hence 4 complex or 8 real scalar fields. (The T.B. models contain two additional real scalar fields from the U multiplet.) Three of these states are massless Goldstone bosons absorbed by the $W^\pm$ and Z^0 bosons, leaving 5 (or 5+2) massive real modes. The general analysis is somewhat complex, and we summarize here some of the more important features.

(i) T.B. Models

The 7 modes rearrange into one charged state of mass

$$m_{H\pm}^2 = M_W^2 + 2(1+\beta_1^2)m_g^2 \quad , \tag{7.26}$$

one neutral state of mass

$$m_{Ho}^2 = M_z^2 + 2(1+\beta_1^2)m_g^2 \tag{7.27}$$

and four additional neutral modes mixed by the couplings λ' and λ'' of Eq. (7.2) [E5,D11]. In Eqs. (7.26), (7.27) β_1 is a model dependent parameter of O(1), and so these Higgs bosons lie above the W and Z bosons [and probably considerably for $m_g \sim O(M_W)$].

(ii) R.G. Models

Here the couplings λ' and λ'' of Eq. (7.2) are zero, and the 5 massive modes arrange themselves into three neutral Higgs mesons H^o, $H^o{}_{(1,2)}$ and one charged meson $H^{\pm}$ with masses given by [E12]

$$m_{H^o}^2 = \frac{mg^2}{\cos^2\alpha}[1 + \mu^2/m_1^2 - \frac{1}{2} \cos 2\alpha \frac{M_z^2}{m_g^2}] \tag{7.28}$$

$$m_{H^o{}_{(1,2)}}^2 = \frac{1}{2}[(M_z^2 + m_{H^o}^2) \pm \{(M_z^2 + m_{H^o}^2)^2 - 4(\cos 2\alpha)^2 m_{H^o}^2 M_z^2\}^{1/2}] \tag{7.29}$$

$$m_{H+}^{2} = M_{W}^{2} + m_{H^{0}}^{2} \quad . \tag{7.30}$$

(These expressions reduce to the results in [D17] in the limit $\alpha \to 0$.) We note that Eq. (7.28) requires

$$m_{g}^{2} > \frac{1}{2} \cos 2\alpha \, M_{z}^{2} \approx (60\ \mathrm{GeV})^{2} \tag{7.31}$$

to prevent the H^{0} mode from becoming tachyonic. However, if m_g is not large, it is possible for H^{0} and also $H^{0}{}_{(2)}$ to be quite lowlying. Such neutral Higgs bosons if produced would decay into hadron and lepton pairs and hence might be detectable at current accellerators. The remaining two Higgs bosons, H^{+} and $H^{0}{}_{(1)}$ lie above the W and Z bosons respectively.

VIII. SUPERSYMMETRIC DECAY OF $W^{\pm}$ AND Z^{0} BOSONS

The $W^{\pm}$ and Z^{0} bosons interact with the gauginos and Higgsinos by standard SU(2)XU(1) gauge interactions. Having found the fields representing the physical particles of the theory in Sec. VII, one may eliminate the elementary fields in terms of them and calculate the vertices for the physical decays of the $W^{\pm}$ and Z^{0} particles. There are four interesting supersymmetric decays of these vector bosons.

(i) $W^{\pm} \rightarrow \tilde{W}_{-}^{\pm} + \tilde{\gamma}$ [E4,E8,E9].

This decay is energetically possible provided

$$\tilde{m}_{-} + \tilde{m}_{\gamma} < M_W \tag{8.1}$$

and as we have seen from Eq. (7.16) the lower Wino mass $\tilde{m}_{-}$ obeys $\tilde{m}_{-} < M_W$ almost always, and so this decay can occur in almost all models. The interaction governing the decay is

$$L_{W\tilde{W}\tilde{\gamma}} = - e\ \bar{\lambda}^{\gamma}\gamma^{\mu}[\sin\gamma_{+}P_{+} - \cos\gamma_{-}P_{-}]\tilde{W}_{-}W_{\mu}^{\dagger} + \text{h.c.} \tag{8.2}$$

where $P_{\pm} = (1/2)(1\pm\gamma_5)$ and $\gamma_{\pm}$ is given in Eq. (7.15)

(ii) $Z^{0} \rightarrow \tilde{W}^{+}_{-} + \tilde{W}^{-}_{-}$ [E5,E9,E12]

This mode requires

$$2\tilde{m}_- < M_Z \tag{8.3}$$

and since a light Wino is expected in all models, it is energetically feasible in all models. The vertex interaction governing the decay is

$$L_{Z\widetilde{W}\widetilde{W}} = - e\, \widetilde{W}_-\gamma^\mu[A_+P_+ + A_-P_-]\widetilde{W}_-Z_\mu \tag{8.4}$$

where

$$A_+ = \cot\theta_W \sin^2\gamma_+ + \cot 2\theta_W \cos^2\gamma_+$$

$$A_- = \cot\theta_W \cos^2\gamma_- + \cot 2\theta_W \sin^2\gamma_- \quad . \tag{8.5}$$

(iii) $W^\pm \to \widetilde{W}^\pm_- + \widetilde{Z}^o_-$ [E9,E12]

This mode requires

$$\tilde{m}_- + \tilde{\mu}_- < M_W \tag{8.6}$$

where the Wino and Zino masses $\tilde{m}_-$, $\tilde{\mu}_-$ are given in Eqs. (7.13) and (7.21a). The mode is feasible only in the T.B. model, for as discussed in Sec. VII only there can the $\widetilde{Z}_-$ be light. For $\alpha = 45^o$, the decay vertex is given by

$$L_{W\tilde{W}\tilde{Z}} = \frac{ie}{\sin\theta_W} A \,\bar{\tilde{Z}}_{-}\gamma^{\mu}\tilde{W}_{-}W_{\mu}^{+} + \text{h.c.} \tag{8.7}$$

where

$$A_W = \cos\theta \,\sin(\beta_+ + \frac{\pi}{4})0_{12} + \frac{1}{2}\sin\,(\beta_+ - \frac{\pi}{4})0_{22} \tag{8.8}$$

and $0_{ik} = N_{ik}/D_k$ is given in Eq. (7.23).

(iv) $Z^{o} \to \tilde{Z}_{(3)} + \tilde{Z}_{(3)}$ [E9,E12] .

Here one requires a light Twilight Zino with mass

$$2\tilde{\mu}_3 < M_z \tag{8.9}$$

and hence this mode occurs in almost all R.G. models (but is energetically forbidden in T.B. models). The decay vertex here is

$$L_{Z\tilde{Z}_3\tilde{Z}_3} = -\frac{e}{2\sin2\theta_W}[\cos2\alpha\{(0_{23})^2 - (0_{33})^2\}$$

$$-2\sin2\alpha 0_{23}0_{33}]\bar{\tilde{Z}}_{(3)}\gamma^{\mu}\gamma_5\tilde{Z}_3 Z_{\mu} \quad . \tag{8.10}$$

The above decay interactions depend only on the mixing angle α of Eq. (7.3), the paramter μ of Eq. (7.2) and the photino mass $\tilde{m}_{\gamma}$. For the R.G. model, α and μ determine the Wino and Twilight masses ($\tilde{m}_{-}$ and $\tilde{\mu}_3$) while in the T.B. model $\alpha \cong 45^{o}$ and μ determines $\tilde{m}_{-}$. Thus for fixed choices of $\tilde{m}_{-}$, $\tilde{\mu}_3$ and $\tilde{m}_{\gamma}$ one obtains unique predictions for the decay rates. Characteristic results are given in Table 1. The remarkable feature is the largeness of the supersymmetric branching ratios, particularly the $Z^0 \to \tilde{W}^+ + \tilde{W}^-$ which are of "industrial strength" size in all models! In order to see what experimental signals these decays give, it is necessary first, however, to examine the Wino and Zino decay modes.

The $\tilde{W}_-$ decays proceed through intermediate squark, selectron and W states. The diagrams governing $\tilde{W}_-$ decays are shown in Fig. 3 with the following decay processes possible [E9,E12]:

$$\tilde{W}_-^+ \to u_i + \bar{d}_i + \tilde{g}$$

$$\tilde{W}_-^+ \to u_i + \bar{d}_i + \tilde{\gamma}$$

$$\tilde{W}_-^+ \to l^+ + \nu_l + \tilde{\gamma} \quad . \tag{8.11}$$

Here $\tilde{g}$ = gluino, l^+ = lepton and u_i and d_i stand for up and down type quark (i is a genertion index.) The interactions governing the vertices can be calculated using Eqs. (7.7) and (7.8). Thus the Wino-quark-squark vertex

$$L_{\tilde{W}q\tilde{q}} = \frac{-ie}{\sqrt{2}\sin\theta_W} [-\sin\gamma_+ \bar{u} P_+ \tilde{W}_- \tilde{d}_L + \cos\gamma_- \bar{d} P_+ \tilde{W}_-^c \tilde{u}_L] + \text{h.c.} \tag{8.12}$$

and the gluino-quark-squark vertex is

$$L_{\tilde{g}q\tilde{q}} = ig_3 (\frac{t^r}{2})_{ij} [\bar{u}_i P_+ \lambda_r \tilde{u}_{jL} + \bar{d}_i P_+ \lambda_r \tilde{d}_{jL}] + \tag{8.13}$$
$$+ ig_3 (t^r/2)_{ij} [\bar{u}_i P_- \lambda_r \tilde{u}_{jR} + \bar{d}_i P_- \lambda_r \tilde{d}_{jL}] + \text{h.c.}$$

where $\lambda_r(x)$ is the Majorana gluino field and t^r the SU(3) matrices. In general there is a significant amount of interference between the W and squark (and W and selectron) poles which must be taken into account, and the gluino modes are strongly reduced due to the gluino mass in the three body phase space.

A similar set of squark and selectron poles lead to the following Zino decay modes (See Fig. 6):

$$\tilde{Z}_- \rightarrow u_i(d_i) + \bar{u}_i(\bar{d}_i) + \tilde{g}$$

$$\tilde{Z}_- \rightarrow u_i(d_i) + \bar{u}_i(\bar{d}_i) + \tilde{\gamma}$$

$$\tilde{Z}_- \rightarrow l^+ + l^- + \tilde{\gamma} \; . \qquad (8.14a)$$

In addition, the $W^{\pm}$ pole diagrams can lead to $\tilde{W}^{\pm}$ final states, since the $\tilde{W}_-$is lighter than the $\tilde{Z}_-$:

$$\tilde{Z}_- \rightarrow u_i(l^+) + \bar{d}_i(u_l) + \tilde{W}_-^-$$

$$\tilde{Z} \rightarrow \bar{u}_i(l^-) + d_i(\bar{u}_l) + \tilde{W}_-^+ \; . \qquad (8.14b)$$

Finally we note that the Twilight zino decay is via the squark and selectron intermediate states:

$$\tilde{Z}_{(3)} \rightarrow u_i(d_i) + \bar{u}_i(\bar{d}_i) + \tilde{\gamma}$$

$$\tilde{Z}_{(3)} \rightarrow l^+ + l^- + \tilde{\gamma} \quad . \tag{8.15}$$

The decay branching ratios for the $\tilde{W}_-$, $\tilde{Z}_-$ and $\tilde{Z}_3$ are given in Tables 2 and 3 for two values of photino mass $\tilde{m}_\gamma = 2$ and $\tilde{m}_\gamma = 7$ GeV. From Eq. (1.3) one sees that the gluino final states are energetically forbidden for the heavier photino choice, reducing some of the hadronic branching ratios.

Combining Table 1 with Tables 2 and 3 leads to the branching ratios given in Tables 4 and 5 for various final states in the supersymmetric decays of the $W^\pm$ and Z^0 bosons. Again we note the largeness of some of the decay modes, particularly in the Z^0 decays. There are a number of significant exerimental signals, some of which may be detectable now at the CERN $\bar{p}p$ Collider or at the e^+e^- SLC and LEP machines. These fall into the following general catagories

(a) UFO Events

These are $W^\pm$ and Z^0 decays into 1 or 2 jets with unbalanced high $p_\perp$ where "unidentified fermionic objects" (UFOs), i.e. the photinos, take away missing $p_\perp$ with no additional leptons present. Events of this type with one relatively broad jet Fig. 4a (arising from two quarks in a relatively slow $\tilde{W}$ and $\tilde{Z}$ hadronic decay) come from $W \rightarrow (\tilde{W} \rightarrow \text{jet} + \tilde{\gamma}) + \tilde{\gamma}$, [E4,E8,E9] while events with two broad jets Fig. 4b in opposite hemispheres come from $W \rightarrow \tilde{W}+\tilde{Z}$ and $Z \rightarrow \tilde{W}+\tilde{W}$ decays where the Winos and Zinos decay hadronically [E9]. One expects that most of these events will have a large missing $p_\perp$. We see from Tables 4 and 5 that UFO events occur with sizable probability even at the CERN $\bar{p}p$ Collider in all models and will be particularly prominant at LEP and SLC.

The two jet UFO events given the stronger of the two signals (particularly for the T.B. models). At the $\bar{p}p$ Collider 2 jet events possess one possible background, i.e. gluino pair production $p+\bar{p} \to \tilde{g}+\tilde{g}$, though this process becomes negligible at CERN for gluino masses greater than about 70–80 GeV, See e.g. [F5] pg. 252. (In any event the discovery of a gluino or Wino would be equally exciting!)

(b) Lepton-Jet Events With Missing $p_{\perp}$

These events arise from $W \to \tilde{W} + \tilde{Z}$ or $Z \to \tilde{W} + \tilde{W}$ where one "ino" decays hadronically and one leptonically. [E9]. One expects, therefore a lepton in one hemisphere, and a broad jet in the other with missing $p_{\perp}$ (Fig. 8). We estimate that 30–40% of these events will satisfy the experimental $p_{\perp elec} > 15$ GeV cut, and in the T.B. model roughly at a rate ≈15% of the $W \to e\nu$ events at CERN. A possible background for this process would be heavy flavor production e.g. top quark pairs [F6]. However, Fig. 5 differs from the t-quark signature in that there will be no b quark decay debris, and there should be additional missing $p_{\perp}$ from the $\tilde{\gamma}$ arising in the hadronic decay. We note also that the appearance of events such as Fig. 5 at the $\bar{p}p$ Collider would argue in favor of the T.B. models, though lepton-jet events should occur in all models at SLC and LEP from Z^{0} decays.

(c) Exotic Leptonic Decays With Missing $p_{\perp}$

As can be seen from Tables 4 and 5, supersymmetry predicts a number of purely leptonic exotic W and Z decays with small but non-negligible branching ratios [E9]. Thus all models predict a sizable decay rate for $Z \to$

$l^+{}_1 + l^-{}_2$ + neutrals (where l_1 and l_2 may be in different families). The R.G. models predict $Z \to l^+{}_1 + l^-{}_1 + l^+{}_2 + l^-{}_2$ + neutrals (from the Twilight decays) while the T.B. models predict $W^+ \to l^+{}_1 + l^+{}_2 + l^-{}_2$ + neutrals. If exotic decays such as these exist, they could presumably be observed at LEP or SLC.

(d) The low energy effective Lagrangian contains interactions of W and Z with sleptons. Whether or not the corresponding decays of the W and Z occur depends on the slepton mass spectra [E25,E26]. Similar mass spectra considerations hold for the possible generation of squarks and sleptons in other processes [E27,E28].

	Branching Ratio (fraction)			
	Tree Breaking		Renormalization Group	
Decay	$\tilde{m}_\gamma$ = 2GeV	$\tilde{m}_\gamma$ = 7GeV	$\tilde{m}_\gamma$ = 2GeV	$\tilde{m}_\gamma$ = 7GeV
$W \to \tilde{W} + \tilde{\gamma}$	.57	.48	.35	.33
$W \to \tilde{W} + \tilde{Z}$	1.88	1.63	0	0
$Z^o \to \tilde{W}^+ + \tilde{W}^-$	7.14	6.47	4.53	4.45
$Z^o \to \tilde{Z}_{(3)} + \tilde{Z}_{(3)}$	0	0	1.31	1.32

Table 1. Branching ratios for the supersymmetric decays of the W and Z bosons (relative to $W \to e\nu$ and $Z \to e^+e^-$ respectively) for the tree breaking model and renormalization group model. Values quoted are for Wino mass of 30 GeV, $m_{\tilde{g}} = \sqrt{2}M_W$ and $\tilde{Z}_{(3)}$ mass - 9.5 GeV for two values of the photino mass $\tilde{m}_\gamma$. Results of this table and of the following tables are taken from Ref. [E9].

	Branching Ratio (%)	
	$\tilde{m}_\gamma = 2$ GeV	$\tilde{m}_\gamma = 7$ GeV
$\tilde{W} \rightarrow 1 + \nu_1 + \tilde{\gamma}$	21.8	26.2
$\tilde{W} \rightarrow h + \tilde{\gamma}$	78.2	73.8
$\tilde{Z} \rightarrow 1 + \nu_1 + \tilde{W}$	0.4	11.5
$\tilde{Z} \rightarrow h + \tilde{W}$	1.2	33.0
$\tilde{Z} \rightarrow 1 + \bar{1} + \tilde{\gamma}$	1.9	16.0
$\tilde{Z} \rightarrow h + \tilde{\gamma}$	96.0	36.1

Table 2. Branching ratios for Wino and Zino decays for the tree breaking model for two photino masses. 1 means e or μ leptons and h means hadrons. The analysis is for $m_g = \sqrt{2}M_W$.

	Branching Ratio (%)	
	$\tilde{m}_\gamma = 2$ GeV	$\tilde{m}_\gamma = 7$ GeV
$\tilde{W} \rightarrow 1 + \nu_1 + \tilde{\gamma}$	17.6	20.9
$\tilde{W} \rightarrow h + \tilde{\gamma}$	82.4	79.1
$\tilde{Z}_{(3)} \rightarrow 1^+ + 1^- + \tilde{\gamma}$	21.6	34.7
$\tilde{Z}_{(3)} \rightarrow h + \tilde{\gamma}$	73.8	65.3
$\tilde{Z}_{(3)} \rightarrow h + 1 + \nu_1 + \tilde{\gamma}$	4.6	0

Table 3. Branching ratios for Wino and "Twilight Zino" ($Z_{(3)}$) for renormalization group model for two photino masses. 1 means e or μ leptons and h means hadrons. Wino decays are for $m_g = 150$ GeV and the $\tilde{Z}_{(3)}$ decays are for $m_g = \sqrt{2}M_W$.

Decay	Branching Ratio (%)	
	$\tilde{m}_\gamma$ = 2 GeV	$\tilde{m}_\gamma$ = 7 GeV
$W \to \gamma + (\tilde{W} \to h + \gamma)$	44.6	35.1
$W \to \gamma + (\tilde{W} \to l + \nu_1 + \tilde{\gamma})$	12.5	12.5
$W \to \tilde{W} + \tilde{Z} \to (l + \nu_1 + \tilde{\gamma}) + (h + \tilde{\gamma})$	41.2	50.6
$W \to (\tilde{W} \to l_1 + \nu_1 + \tilde{\gamma}) + (\tilde{Z} \to l_2 + \bar{l}_2 + \tilde{\gamma})$	0.8	8.2
$W \to (\tilde{W} \to h + \gamma) + (\tilde{Z} \to l + \bar{l} + \tilde{\gamma})$	2.9	19.3
$W \to (\tilde{W} \to h + \tilde{\gamma}) + (\tilde{Z} \to h + \gamma)$	142.1	72.8
$Z \to (\tilde{W} \to l + \nu_1 + \tilde{\gamma}) + (\tilde{W} \to h + \tilde{\gamma})$	243.7	250.6
$Z \to (\tilde{W} \to l_1 + \nu_1 + \tilde{\gamma}) + (\tilde{W} \to l_2 + \nu_2 + \tilde{\gamma})$	34.0	44.6
$Z \to (\tilde{W} \to h + \gamma) + (\tilde{W} \to h + \tilde{\gamma})$	436.4	352.2

Table 4. Branching ratios for W decays relative to $W \to e + \nu$ and Z decays relative to $Z \to e^+ + e^-$ for the tree breaking model. l, l_1, l_2 stand for e or μ leptons and h for hadrons. The analysis is for $m_g = \sqrt{2}M_W$.

Decay	Branching Ratio (%) $\tilde{m}_\gamma$ = 2 GeV	$\tilde{m}_\gamma$ = 7 GeV
$W \to \tilde{\gamma}\ (\tilde{W} \to h + \tilde{\gamma})$	29.1	26.2
$W \to \tilde{\gamma} + (\tilde{W} \to l + \nu_1 + \tilde{\gamma})$	6.2	6.9
$Z \to (\tilde{W} \to l + \nu_1 + \tilde{\gamma}) + (\tilde{W} \to h + \tilde{\gamma})$	131.0	147.0
$Z \to (\tilde{W} \to l_1 + \nu_1 + \tilde{\gamma}) + (l_2 + \nu_2 + \tilde{\gamma})$	13.9	19.4
$Z \to (\tilde{W} \to h + \tilde{\gamma}) + (\tilde{W} \to h + \tilde{\gamma})$	307.6	278.4
$Z \to (\tilde{Z}_{(3)} \to l_1 + \bar{l}_1 + \tilde{\gamma}) + (\tilde{Z}_{(3)} \to l_2 + \bar{l}_2 + \tilde{\gamma})$	6.1	15.8
$Z \to (\tilde{Z}_{(3)} \to l + \bar{l} + \tilde{\gamma}) + (\tilde{Z}_{(3)} \to h + \tilde{\gamma})$	41.9	59.6
$Z \to (\tilde{Z}_{(3)} \to h + \tilde{\gamma}) + (\tilde{Z}_{(3)} \to h + \tilde{\gamma})$	71.5	56.2
$Z \to (\tilde{Z}_{(3)} \to h + \tilde{\gamma}) + (\tilde{Z}_{(3)} \to h + l + \nu_1 + \tilde{\gamma})$	8.8	0

Table 5. Branching ratios for W decays relative to $W \to e + \nu$ and Z decays relative to $Z \to e^+ + e^-$ for renormalization group model. l, l_1 l_2 stand for e or μ leptons and h for hadrons. Branching ratios through the Wino poles are for m_g = 150 GeV and through the $\tilde{Z}_{(3)}$ poles are for $m_g = \sqrt{2}M_W$.

IX. CONCLUSION

N=1 Supergravity unified models generate a dynamical unification of electro-weak and supergravitational interactions. There are a large number of predictions of such a unification at low energy. The theory predicts an array of new particles, photino, gluino, winos, Zinos, Higgsinos, sleptons, and squarks, with characteristic mass scales governed by the gravitino mass mg ~ $O(m_W)$. Of these the lighest particles are expected to be the photino, the wino below the W boson and the Zino below the Z boson and hence they represent the best chance of being discovered at current energies at the $\bar{p}p$ collider. The best chance for discovering the selectron and the sneutrino would be at LEP or SLC.

A number of other theoretical consequences of N=1 Supergravity unified models have also been investigated recently. These models suggest possible additional sources of CP violation and could generate contributions to the electric dipole moment of the neutron and the electron which are close to the current experimental upper bounds [E29]. Further, the recent experimental lower limits on the proton decay [F7] require an assessment of the conventional grand unification program [F7], while further theoretical analysis of Supergravity GUT predictions for the strange decay modes is needed. Finally N=1 supergravity models appear to fare better than the ordinary GUT or globally supersymmetric models in allowing for acceptable inflationary scenarios of the early universe [E30].

ACKNOWLEDGEMENTS

This work is supported in part by the National Science Foundtion under Grants No. PHY77-22864 and No. PHY80-08333. One of us (P.N.) wishes to thank the International Center for Theoretical Physics (ICTP) at Trieste for the hospitality accorded him during the period of his visit there.

APPENDIX A: CONSTRUCTION OF SUPERGRAVITY-MATTER LAGRANGIAN

In this appendix we shall explain in detail the steps needed to construct the Lagrangian of locally supersymmetric grand unified theories [C8].

We consider the coupling of supergravity with the minimal set of auxiliary fields to the gauge vector multiplet and to an arbitrary number of scalar multiplets, (which are representations of the gauge group) and where the full Langrangian must be locally supersymmetric, and locally gauge invariant. For simplicity we assume Eq. (2.21) to hold. The final result can be obtained by different equivalent methods. Our analysis here will be carried in terms of the component fields of the supermultiplets, using the rules of tensor calculus for chiral [C3] and vector multiplets [C2]. The supergravity Lagrangian [C1] is given in Eq. (2.1) where the fields A_μ and u are auxiliary.

This Lagrangian is invarient under the following supersymmetry transformations:

$$\delta_s e^r_\mu = \kappa\bar{\varepsilon}\gamma^r\psi_\mu$$

$$\delta_s\psi_\mu = 2\kappa^{-1}D_\mu[\omega(e,\psi)]\varepsilon + i\gamma_5(\delta^\nu_\mu - \tfrac{1}{3}\gamma_\mu\gamma^\nu)\varepsilon A_\nu + \tfrac{1}{3}\gamma_\mu(S - i\gamma_5 P)\varepsilon$$

$$\delta_s S = \tfrac{1}{2}e^{-1}\bar{\varepsilon}\gamma_\mu R^\mu + i\tfrac{\kappa}{2}\bar{\varepsilon}\gamma_5\psi_\nu A^\nu - \tfrac{\kappa}{2}\bar{\varepsilon}(S + i\gamma_5 P)\gamma^\mu\psi_\mu$$

$$\delta_s P = -\tfrac{i}{2}e^{-1}\bar{\varepsilon}\gamma_5\gamma_\mu R^\mu + \tfrac{\kappa}{2}\bar{\varepsilon}\psi_\nu A^\nu + i\tfrac{\kappa}{2}\bar{\varepsilon}\gamma_5(S + i\gamma_5 P)\gamma^\mu\psi_\mu \tag{A.1}$$

$$\delta_s A_\mu = 3\tfrac{i}{2}e^{-1}\bar{\varepsilon}\gamma_5(\delta^\nu_\mu - \tfrac{1}{3}\gamma_\mu\gamma^\nu)R_\nu + \kappa\bar{\varepsilon}\gamma^\nu(\psi_\mu A_\nu - \tfrac{1}{2}\psi_\nu A_\mu)$$

$$- e\tfrac{\kappa}{4}\varepsilon_{\mu\nu\rho\sigma}\bar{\varepsilon}\gamma_5\gamma^\rho\psi^\sigma A^\nu + i\tfrac{\kappa}{2}\bar{\varepsilon}\gamma_5(S - i\gamma_5 P)\psi_\mu$$

where $\varepsilon(x)$ is the supergravity parameter. The Lagrangian for the vector multiplet (defined in Eq. (2.10)) in the Wess-Zumino gauge, and with a normalize kinetic energy, when coupled to supergravity is given in Eq. (2.20). The expression appearing in Eq. (2.20) are defined by

$$V = V^{\alpha}T^{\alpha} = (V_{\mu}^{\alpha}, \lambda^{\alpha}, D^{\alpha})T^{\alpha}$$

$$[T^{\alpha}, T^{\beta}] = 2if^{\alpha\beta\gamma}T^{\gamma} \tag{A.2}$$

$$F_{\mu\nu}{}^{\alpha} = \partial_{\mu}V_{\nu}^{\alpha} - \partial_{\nu}V_{\mu}^{\alpha} + g_{\alpha}f^{\alpha\beta\gamma}V_{\mu}^{\beta}V_{\nu}^{\gamma}$$

$$D_{\mu}\lambda^{\alpha} = \partial_{\mu}\lambda^{\alpha} + g_{\alpha}f^{\alpha\beta\gamma}V_{\mu}^{\beta}\lambda^{\gamma} + \frac{1}{2}\omega_{\mu rs}\sigma^{rs}\lambda^{\alpha} + i\frac{\kappa}{2}A_{\mu}\gamma_5\lambda^{\alpha}; \quad \sigma^{rs} \equiv \frac{1}{4}[\gamma^{r}, \gamma^{s}]$$

where g_{α} is the gauge group coupling constant, and $T^{\alpha}/2$ are the gauge group generators (The T^{α} here is normalized, similarly to Pauli-matrices). The Lagrangian of Eq. (2.20) is invariant under the supersymmetry transformations:

$$\delta_s V_{\mu}^{\alpha} = \bar{\varepsilon}\gamma_{\mu}\lambda^{\alpha}$$

$$\delta_s\lambda^{\alpha} = -\sigma^{\mu\nu}\bar{\varepsilon}\hat{F}_{\mu\nu}^{\alpha} - i\gamma_5\varepsilon D^{\alpha} \tag{A.3}$$

$$\delta_s D^{\alpha} = -i\bar{\varepsilon}\gamma_5\hat{\not{D}}\lambda^{\alpha} .$$

The supercovarient quantities $\hat{F}_{\mu\nu}{}^{\alpha}$ and $\hat{D}_{\mu}\lambda^{\alpha}$ that appears in (A.3) are related to $F_{\mu\nu}{}^{\alpha}$ and $D_{\mu}\lambda^{\alpha}$ defined in (A.2) by

$$\hat{F}_{\mu\nu}{}^{\alpha} = F_{\mu\nu}{}^{\alpha} - \frac{\kappa}{2}(\bar{\psi}_{\mu}\gamma_{\nu}\lambda^{\alpha} - \bar{\psi}_{\nu}\gamma_{\mu}\lambda^{\alpha})$$

$$\hat{D}_\mu \lambda^\alpha = D_\mu \lambda^\alpha + \frac{\kappa}{2}(\sigma^{\mu\nu}\hat{F}_{\mu\nu}{}^\alpha + i\gamma_5 D^\alpha)\psi_\mu \ . \tag{A.4}$$

The left-handed chiral multiplets Σ^a are defined in Eq. (2.9), and under supersymmetry transformations the component fields transforms as follows:

$$\delta_s z^a = 2\bar{\varepsilon}_R \chi^a$$

$$\delta_s \chi^a = h^a \varepsilon_L + \hat{\not{D}} z^a \varepsilon_R$$

$$\delta_s h^a = 2\bar{\varepsilon}_R \hat{\not{D}} \chi^a - 2\kappa \bar{\eta}_R \chi^a \tag{A.5}$$

where

$$\eta_R = \frac{1}{3}(u^* \varepsilon_R - i\not{A}\varepsilon_L)$$

$$\hat{D}_\mu z^a = \partial_\mu z^a - \kappa \bar{\psi}_\mu \chi^a$$

$$\hat{D}_\mu \chi^a = [D_\mu \omega(e,\psi) - i\frac{\kappa}{2}A_\mu]\chi^a - \frac{\kappa}{2}\not{D} z^a \psi_{\mu R} - \frac{\kappa}{2} h^a \psi_{\mu L} \ . \tag{A.6}$$

From the multiplet Σ^a we can construct a multiplet of opposite chirality, denoted by

$$\Sigma_a \equiv (\Sigma^a)^+ = (z_a, \chi_a, h_a) \tag{A.7}$$

where

$$z_a = z^{a\dagger} = A^a - iB^a \tag{A.8}$$

$$\chi_a = C^{-1}\overline{\chi^a} \equiv (\chi^a)^c = \text{Right-handed Weyl spinor} \tag{A.9}$$

$$h_a = h^{a\dagger} = F^a - iG^a \tag{A.10}$$

In (A.9), C is the charge conjugation matrix, and in what follows upper and lower indices will be used with left and right handed multiplets respectively. To construct a gauge invariant interaction we need the rules of multiplets multiplication, which are,

(1) Two multiplets of the same chirality when multiplied form a multiplet of the same chirality:

$$\Sigma_1 \cdot \Sigma_2 = (Z_1, \chi_1, h_1) \cdot (Z_2, \chi_2, h_2) = (Z_1 Z_2,\ Z_1\chi_2 + Z_2\chi_1,\ Z_1 h_2$$

$$+ Z_2 h_1 - 2\bar{\chi}_1^c \chi_2) \tag{A.11}$$

this rule, for chiral multiplets with Weyl-spinors, can be obtained directly from superfield multiplication as given by Salam and Strathdee [A2], or indirectly by using the Wess-Zumino rules for multiplets with Majorana spinors [A1]. In the later case we first write the rule for multiplying multiplets with left-handed Majorana spinors (which reads exactly as in (A.11) but with $\bar{\chi}_1^c\chi_2$ replaced by $\bar{\chi}_1\chi_{2L}$) then from each two independent multiplets with Majorana spinors, one multiplet with a Weyl-spinor is formed and rule (A.11) is deduced.

(2) Two multiplets of opposite chiralities, when multipleid symmetrically, give rise to a vector multiplet:

$$\Sigma_1 \times \Sigma_2 \equiv \tfrac{1}{2}(\Sigma_1^\dagger \Sigma_2 + \Sigma_2^\dagger \Sigma_1)$$

$$= (\tfrac{1}{2} Z_1^* Z_2, i(Z_1^*\chi_2 - Z_1\chi_2^c), - h_1 Z_2^*, \tfrac{i}{2}(Z_1^*\hat{D}_\mu Z_2 - Z_1\hat{D}_\mu Z_2^* - 2\bar{\chi}_1\gamma_\mu\chi_2,$$

$$- ih_1^* \chi_2 + ih_1 \chi_2^C - i\hat{\not{D}} Z_1^* \chi_2 + i\hat{\not{D}} Z_1 \chi_2^c ,$$

$$h_1^* h_2 - \hat{D}_\mu Z_1^* \hat{D}_\mu Z_2 - \bar{\chi}_1 \overset{\leftrightarrow}{\hat{\not{D}}} \chi_2) + 1 \leftrightarrow 2 \qquad \text{(A.12)}$$

Note that the spinor components ξ and λ of the multiplet in (A.12) are both Majorana as is required for vector multiplets. To derive Eq. (A.12) we start with the symmetric product rule for multiplets with Majorana spinors as given by Stelle and West C2 (this reads exactly as Eq. (A.12) but with χ^c replaced with χ), then as before, from each two independent multiplets with left-handed Majorana spinors, one multiplet with a Weyl spinor is formed. However, in this case the anti-symmetric product rule for two multiplets of opposite chiralities, with Majorana spinors, is also needed. To see this let Λ_1, Λ_2, Λ_1', Λ_2' be the left-handed Majorana multiplets, and

$$\Sigma_1 = \Lambda_1 + i\Lambda_2$$

$$\Sigma_2 = \Lambda_1' + i\Lambda_2' \qquad \text{(A.13)}$$

then

$$\tfrac{1}{2}(\Sigma_1 \Sigma_2^\dagger + \Sigma_2 \Sigma_1^\dagger) = (\Lambda_1 \Lambda_1' + \Lambda_2 \Lambda_2') - i(\Lambda_1 \Lambda_2' - \Lambda_1' \Lambda_2) .$$

The antisymmetric product rule for multiplets with Majorana spinors reads

$$\Lambda_1 \wedge \Lambda_2 \equiv - \tfrac{i}{2}(\Lambda_1^\dagger \Lambda_2 - \Lambda_1 \Lambda_2^\dagger)$$

$$= (\tfrac{i}{2} Z_1^* Z_2, Z_1^* \chi_{2L} + Z_1 \chi_{2R}, iZ_1^* h_2, \tfrac{1}{2}(Z_1^* D_\mu Z_2 + Z_1^* D_\mu Z_2 - \bar{\chi}_{1L} \gamma_\mu \chi_{2L}),$$

$$h_2^* \chi_{1L} + \hat{\not{D}} Z_2^* \chi_{1L}, i(h_1^* h_2 - \hat{D}_\mu Z_1^* \hat{D}_\mu Z_2 - \bar{\chi}_2 \hat{\not{D}} \chi_1)) - 1 \leftrightarrow 2 \qquad \text{(A.14)}$$

(3) The product of two vector multiplets is a vector multiplet:

$$V_1 \cdot V_2 = (C_1, \xi_1, v_1, V_{\mu 1}, \lambda_1, D_1) \cdot (C_2, \xi_2, v_2, V_{\mu 2}, \lambda_2, D_2)$$

$$= (\tfrac{1}{2}C_1C_2, C_1\xi_2, C_1 v_2 - \tfrac{1}{2}\bar{\xi}_{1R}\xi_{2L}, C_1 V_{\mu 2} - \tfrac{i}{4}\bar{\xi}_1\gamma_\mu\gamma_5\xi_2 ,$$

$$C_1\lambda_2 - \tfrac{1}{2}\hat{\not{D}}C_1\xi_2 + \tfrac{1}{2}\hat{v}_1^*\xi_2 + \tfrac{i}{2}\gamma_5\gamma^\mu\xi_2 V_{\mu 1} , \tag{A.15}$$

$$C_1D_2 - \tfrac{1}{2}\hat{D}_\mu C_1\hat{D}^\mu C_2 - \tfrac{1}{2}V_{\mu 1}V_2^\mu + \tfrac{1}{2}v_1^* v_2 - \bar{\xi}_1\lambda_2 - \tfrac{1}{2}\bar{\xi}_2\hat{\not{D}}\xi_1) + 1 \leftrightarrow 2$$

where

$$\hat{v} = K + i\gamma_5 H$$

$$\hat{D}_\mu C = \partial_\mu C + i\tfrac{\kappa}{2}\bar{\psi}_\mu\gamma_5\xi \tag{A.16}$$

$$\hat{D}_\mu\xi = D_\mu(\omega(e,\psi))\xi - \tfrac{\kappa}{2}(\not{V} + i\gamma_5\hat{\not{D}}C - i\gamma_5\hat{v})\psi_\mu - i\tfrac{\kappa}{2}A_\mu\gamma_5\xi .$$

By applying rule (A.15) we can evaluate the component form of the multiplet $\exp(g_\alpha V)$ [C2]:

$$\exp(g_\alpha V) = \exp(g_\alpha C)(1, g_\alpha\xi, g_\alpha\hat{v} - \frac{g_\alpha^2}{2}\bar{\xi}_{1R}\xi_{2L}, g_\alpha V_\mu - \frac{g_\alpha^2}{4}\bar{\xi}\gamma_5\gamma_\mu\xi,$$

$$g_\alpha\lambda - \frac{g_\alpha^2}{2}\hat{\not{D}}C\xi + \frac{g_\alpha^2}{2}(\hat{v}^* + i\gamma_5\not{V})\xi - \frac{g_\alpha^2}{4}(\bar{\xi}\xi)\xi , \tag{A.17}$$

$$g_\alpha D - \frac{g_\alpha^2}{2}\hat{D}_\mu C\hat{D}^\mu C - \frac{g_\alpha^2}{2}V_\mu V^\mu + \frac{g_\alpha^2}{2}|v|^2 - g_\alpha^2\bar{\xi}\lambda - \tfrac{1}{2}g_\alpha^2\bar{\xi}\hat{\not{D}}\xi$$

$$- \frac{g_\alpha^2}{4}\bar{\xi}(\hat{v}^* + i\gamma_5\not{V})\xi - \frac{g_\alpha^2}{16}(\bar{\xi}\xi)^2) .$$

Equation (A.17) simplifies greatly in the Wess-Zumino gauge ($C = \xi = v = 0$):

$$\exp(g_\alpha V) = (1,0,0,g_\alpha V_\mu, g_\alpha \lambda, g_\alpha D - \frac{g_\alpha^2}{2} V_\mu V^\mu) \ . \tag{A.18}$$

Guided by the vector-scalar coupling of global supersymmetry, we form the gauge invariant interaction

$$\frac{1}{4}(\Sigma_{1a} \ \exp \ (g_\alpha V)^a{}_b \Sigma_2^b + 1 \leftrightarrow 2) \tag{A.19}$$

the above expression is gauge invariant because under group transformations we have:

$$\Sigma^a \to \Omega^a{}_b \Sigma^b$$

$$\Sigma_a \to \Sigma_b \Omega^{\dagger b}{}_a \tag{A.20}$$

$$\exp(g_\alpha V)^a{}_b \to (\Omega^{-1})^{\dagger a}{}_c \ \exp(g_\alpha V)^c{}_d (\Omega^{-1})^d{}_b$$

where $\Omega^a{}_b$ is a chiral left-handed multiplet whose components are parameters for the group transformations.

The Lagrangian for the gauge invariant and locally supersymmetric interaction of supergravity to the scalar multiplet, can be obtained by applying Eq. (2.13) to the components of the vector multiplet resulting from Eq. (A.19), after setting Σ_2^a to be equal to Σ_1^a. However, because $L_{S \cdot G}$ appears in the last term of Eq. (2.13), a Weyl-scaling will be needed and that changes the kinetic interactions to non-normalized from. To have at least the choice of normalized kinetic energies we generalize Eq. (A.19) to:

$$\frac{1}{2}(\Sigma_{a_1}\ldots\Sigma_{a_m}(A\ \exp(g_\alpha V))^{a_1\ldots a_m}{}_{b_1\ldots b_n}\Sigma^{b_1}\ldots\Sigma^{b_n} + h\cdot c) \tag{A.21}$$

where $A^{a_1\ldots a_m}{}_{b_1\ldots b_n}$ are the arbitrary coupling constants that appears in Eq. (2.15). The components of $\Sigma^{a_1}\ldots\Sigma^{a_m}$ can be obtained by repeated application of (A.11):

$$(Z^{a_1}\ldots Z^{a_m}, mZ^{(a_1}\ldots Z^{a_{m-1}}\chi^{a_m)}, mZ^{(a_1}\ldots Z^{a_{m-1}}h^{a_m)} - m(m-1)Z^{(a_1}\ldots Z^{a_{m-2}}\bar{\chi}^{a_{m-1}}\chi^{a_m)} \tag{A.22}$$

We form the symmetric product of $\Sigma^{a_1}\ldots\Sigma^{a_m}$ and $\Sigma^{b_1}\ldots\Sigma^{b_n}$, then multiply the resultant vector multiplet with the vector multiplet $(A\ \exp(g_\alpha V))^{a_1\ldots a_m}{}_{b_1\ldots b_n}$. The component form of (A.21) is

$$C = \frac{1}{2}\phi$$

$$\xi = i(\phi,_a\chi^a - \phi,^a\chi_a)$$

$$v = -(\phi,_a h^a - \phi,_{ab}\bar{\chi}_a\chi^b)$$

$$V_\mu = \frac{i}{2}(\phi,_a\mathcal{D}_\mu Z^a - \phi,^a\mathcal{D}_\mu Z_a - 2\phi,^a{}_b\overline{\chi^a}\gamma^\mu\chi^b)$$

$$\lambda = -i\phi,^a{}_b h_a\chi^b + i\phi,^{ab}{}_c(\overline{\chi^a}\chi_b)\chi^c + i\phi,^a{}_b h^b\chi_a$$

$$-i\phi,_{ab}{}^c(\overline{\chi_a}\chi^b)\chi_c - i\phi,^a{}_b\hat{\mathcal{D}}Z_a\chi^b + i\phi,^a{}_b\hat{\mathcal{D}}Z^b\chi_a$$

$$+\frac{g_\alpha}{2}\lambda^\alpha\phi,_a(T^\alpha)^a{}_b Z^b$$

$$D = \phi,^{a}{}_{b} h_{a} h^{b} - \phi,^{ab}{}_{c} \overline{\chi^{a}} \chi_{b} h^{c} - \phi,_{ab}{}^{c} \overline{\chi_{a}} \chi^{b} h_{c}$$

$$+ \phi,_{ab}{}^{cd} \overline{\chi_{a}} \chi^{b} \cdot \overline{\chi^{c}} \chi_{d} - \phi,^{a}{}_{b} \hat{\mathcal{D}}_{\mu} z_{a} \hat{\mathcal{D}}^{\mu} z^{b}$$

$$- \phi,^{a}{}_{b} \overline{\chi^{a}} \overleftrightarrow{\hat{\not{\mathcal{D}}}} \chi^{b} - \phi,^{a}{}_{bc} \overline{\chi^{a}} \gamma^{\mu} \chi^{b} \hat{\mathcal{D}}_{\mu} z^{c}$$

$$- \phi,_{a}{}^{bc} \overline{\chi_{a}} \gamma^{\mu} \chi_{b} \hat{\mathcal{D}}_{\mu} z_{c} + \frac{g_{\alpha}}{2} D^{\alpha} \phi,_{a} (T^{\alpha})^{a}{}_{b} z^{b}$$

$$- i g_{\alpha} \phi,^{a}{}_{b} (\bar{\lambda}^{\alpha} \chi^{b} (T^{\alpha} z)_{a} - \overline{\chi^{a}} \lambda^{\alpha} (T^{\alpha} z)^{b}) \qquad \text{(A.23)}$$

where the function ϕ has been defined in Eq. (2.15) and all script derivatives are the gauge covariant form of the latin ones, e.g.

$$\hat{\mathcal{D}}_{\mu} z^{a} = \hat{D}_{\mu} z^{a} - i \frac{g_{\alpha}}{2} V_{\mu}^{\alpha} (T^{\alpha})^{a}{}_{b} z^{b} \qquad \text{(A.24)}$$

In the above formulas we have distinguished between up and down indices, and the derivatives of ϕ are denoted by:

$$\phi,_{a} = \frac{\partial \phi}{\partial z^{a}}, \quad \phi,^{a} = \frac{\partial \phi}{\partial z_{a}} \quad . \qquad \text{(A.25)}$$

We now apply Eq (2.12) to the multiplet (A.23) to obtain the most general locally supersymmetric gauge invariant Lagrangian (without higher derivatives):

$$e^{-1} L_{D} = \phi,^{a}{}_{b} h_{a} h^{b} - \phi,^{ab}{}_{c} \overline{\chi^{a}} \chi_{b} h^{c} - \phi,_{ab}{}^{c} \overline{\chi_{a}} \chi^{b} h_{c}$$

$$+ \phi,_{ab}{}^{cd} \overline{\chi_{a}} \chi^{b} \overline{\chi^{c}} \chi_{d} - \phi,^{a}{}_{b} \hat{\mathcal{D}}_{\mu} z_{a} \hat{\mathcal{D}}^{\mu} z^{b}$$

$$- \phi,^{a}{}_{b}\overline{\chi^{a}}\overleftrightarrow{\hat{\mathcal{D}}}\chi^{b} - \overline{\chi^{a}}\gamma^{\mu}\chi^{b}(\phi,^{a}{}_{bc}\hat{\mathcal{D}}_{\mu}z^{c} - \phi,^{ac}{}_{b}\hat{\mathcal{D}}_{\mu}z_{c})$$

$$- ig_{\alpha}\phi,^{a}{}_{b}(\bar{\lambda}^{\alpha}\chi^{b}(T^{\alpha}z)_{a} - \overline{\chi^{a}}\lambda^{\alpha}(T^{\alpha}z)^{b})$$

$$+ \frac{g_{\alpha}}{2}D^{\alpha}\phi,_{a}(T^{\alpha}z)^{a} + \frac{\kappa}{2}\phi,^{a}{}_{b}\bar{\psi}_{\mu}\gamma_{\mu}(h_{a}\chi^{b} + h^{b}\chi_{a})$$

$$- \frac{\kappa}{2}(\bar{\psi}_{\mu}\gamma^{\mu}\chi^{c}\overline{\chi^{a}}\chi_{b}\phi,^{ab}{}_{c} - \overline{\chi^{c}}\gamma^{\mu}\psi_{\mu}\bar{\chi}_{a}\chi^{b}\phi,_{ab}{}^{c})$$

$$- \frac{\kappa}{2}\phi,^{a}{}_{b}(\bar{\psi}_{\mu}\gamma^{\mu}\gamma^{\nu}\chi^{b}\hat{\mathcal{D}}_{\nu}z_{a} - \overline{\chi^{a}}\gamma^{\nu}\gamma^{\mu}\psi_{\mu}\hat{\mathcal{D}}_{\nu}z^{b})$$

$$- i\kappa\frac{g_{\alpha}}{2}\bar{\psi}_{\mu}\gamma_{5}\gamma^{\mu}\lambda^{\alpha}\phi,_{a}(T^{\alpha}z)^{a}$$

$$+ \frac{\kappa}{3}[u^{*}(\phi,_{a}h^{a} - \phi,_{ab}\bar{\chi}_{a}\chi^{b}) + u(\phi,^{a}h_{a} - \phi,^{ab}\overline{\chi^{a}}\chi_{b})]$$

$$+ (i\frac{\kappa}{3}A^{\mu} + e^{-1}\frac{\kappa^{2}}{8}\epsilon^{\mu\nu\rho\sigma}\bar{\psi}_{\nu}\gamma_{\rho}\psi_{\sigma})(\phi,_{a}\hat{D}_{\mu}z^{a} - \phi,^{a}\hat{D}_{\mu}z_{a} - 2\phi,^{a}{}_{b}\overline{\chi^{a}}\gamma_{\mu}\chi^{b})$$

$$+ 4\frac{\kappa}{3}(D_{\mu}\bar{\psi}_{\nu}\sigma^{\mu\nu}\chi^{a}\phi,_{a} - \phi,^{a}\overline{\chi^{a}}\sigma^{\mu\nu}D_{\mu}\psi_{\nu})$$

$$- \frac{\kappa^{2}}{3}\phi(-\frac{R}{2\kappa^{2}} - \frac{e^{-1}}{2}\epsilon^{\mu\nu\rho\sigma}\bar{\psi}_{\mu}\gamma_{5}\gamma_{\nu}D_{\rho}\psi_{\sigma} - \frac{1}{3}|u|^{2} + \frac{1}{3}A_{\mu}A^{\mu})$$

$$- \frac{\kappa^{3}}{8}\epsilon^{\mu\nu\rho\sigma}e^{-1}\bar{\psi}_{\mu}\gamma_{\nu}\psi_{\rho}(\bar{\psi}_{\sigma}\chi^{a}\phi,_{a} - \overline{\chi^{a}}\psi_{\sigma}\phi,^{a})\,. \qquad (A.26)$$

All derivatives in (A.26) contain torsion pieces and are gauge covariant. It is still possible to add another piece to the Lagrangian that corresponds to the self interactions of the chiral multiplets. The most general form of such interaction is $g(\Sigma)$ where the function g has been defined in Eq. (2.14). The components of this left-handed chiral multiplet are

$$g(\Sigma^a) = (g(z^a)\ ,\ g,_a\chi^a,\ g,_a h^a - 2g,_{ab}\overline{\chi_a}\chi^b)\ . \tag{A.27}$$

Using Eq. (2.12) on the components of (A.27) we obtain the locally supersymmetric form of this interaction

$$e^{-1}L_{pot} = \frac{1}{2}(g,_a h^a - 2g,_{ab}\overline{\chi_a}\chi^b + \kappa u g + \kappa g,_a\bar{\psi}_\mu\gamma^\mu\chi^a$$
$$+ \kappa^2 g\bar{\psi}_\mu\sigma^{\mu\nu}\psi_{\nu R} + h\cdot c)\ . \tag{A.28}$$

The total Lagrangian is thus

$$L = L_V + L_{pot} + L_D\ . \tag{A.29}$$

The supergravity Lagrangian is already included in L_D and corresponds to the constant part of the function ϕ.

The auxiliary fields u, A_μ, h^a, and D^α appear in (A.29) and must be eliminated to determine the effective interactions of the physical fields. However, because some contributions of the auxiliary fields are buried in the supercovariant derivatives as in Eqs. (A.2), (A.4) and (A.6), we thus expand all these supercovariant derivatives keeping only their torsionless parts

$$e^{-1}L = \frac{\phi}{6}(R(\omega(e)) + e^{-1}\kappa^2\varepsilon^{\mu\nu\rho\sigma}\,\bar{\psi}_\mu\gamma_5\gamma_\nu D_\rho(\omega(e))\psi_\sigma + \frac{2}{3}\kappa^2|u|^2 - \frac{2}{3}\kappa^2 A_\mu A^\mu)$$
$$- \phi,^a{}_b(\mathcal{D}_\mu z_a\mathcal{D}_\mu z^b + \overline{\chi^a}\overleftrightarrow{\not{\mathcal{D}}}(\omega(e))\chi^b - h_a h^b)$$
$$+ \kappa\phi,^a{}_b(\mathcal{D}_\nu z^b\overline{\chi^a}\gamma^\mu\gamma^\nu\psi_\mu + \bar{\psi}_\mu\gamma^\nu\gamma^\mu\chi^b\mathcal{D}_\nu z_a)$$
$$- \overline{\chi^a}\gamma^\mu\chi^b(\phi,^a{}_{bc}\mathcal{D}_\mu z^c - \phi,_b{}^{ac}\mathcal{D}_\mu z_c)$$

$$
\begin{aligned}
&-\phi,{}^{ab}{}_{c}\overline{\chi^{a}}\chi_{b}h^{c} - \phi,{}_{ab}{}^{c}\overline{\chi_{a}}\chi^{b}h_{c} + \frac{g_{\alpha}}{2}D^{\alpha}\phi,{}_{a}(T^{\alpha}Z)^{a} \\
&-ig_{\alpha}\phi,{}^{a}{}_{b}((T^{\alpha}Z)_{a}\bar{\lambda}^{\alpha}\chi^{b} - \overline{\chi^{a}}\lambda^{\alpha}(T^{\alpha}Z)^{b}) \\
&- i\kappa\frac{g_{\alpha}}{4}\bar{\psi}_{\mu}\gamma_{5}\gamma^{\mu}\lambda^{\alpha}\phi,{}_{a}(T^{\alpha}Z)^{a} \\
&+ \frac{\kappa}{3}(u^{*}(\phi,{}_{a}h^{a} - \phi,{}_{ab}\overline{\chi_{a}}\chi^{b} + \frac{3}{2}g^{*}) + u(\phi,{}^{a}h_{a} - \phi,{}^{ab}\overline{\chi^{a}}\chi_{b} + \frac{3}{2}g)) \\
&+ i\frac{\kappa}{3}A^{\mu}(\phi,{}_{a}\mathcal{D}_{\mu}Z^{a} - \phi,{}^{a}\mathcal{D}_{\mu}Z_{a} - \kappa\phi,{}_{a}\bar{\psi}_{\mu}\chi^{a} + \kappa\phi,{}^{a}\overline{\chi^{a}}\psi_{\mu} \\
&+ \phi,{}^{a}{}_{b}\overline{\chi^{a}}\gamma_{\mu}\chi^{b} - \frac{3}{4}\bar{\lambda}^{\alpha}\gamma_{\mu}\gamma_{5}\lambda^{\alpha}) \\
&- e^{-1}\frac{i\kappa^{2}}{8}\varepsilon^{\mu\nu\rho\sigma}\bar{\psi}_{\mu}\gamma_{\nu}\psi_{\rho}(\phi,{}_{a}\mathcal{D}_{\sigma}Z^{a} - \phi,{}^{a}\mathcal{D}_{\sigma}Z_{a}) \\
&+ 4\frac{\kappa}{3}(\overline{D_{\mu}(\omega(e))\psi_{\nu}}\sigma^{\mu\nu}\chi^{a}\phi,{}_{a} - \phi,{}^{a}\overline{\chi^{a}}\sigma^{\mu\nu}D_{\mu}(\omega(e))\psi_{\nu}) \\
&+ \frac{1}{2}(g,{}_{a}h^{a} - g,{}_{ab}\overline{\chi_{a}}\chi^{b} + \kappa g,{}_{a}\bar{\psi}_{\mu}\gamma^{\mu}\chi^{a} + \kappa^{2}g\bar{\psi}_{\mu}\sigma^{\mu\nu}\psi_{\nu R} \\
&+ g,{}^{*a}h_{a} - g,{}^{*ab}\overline{\chi_{a}}\chi_{b} + \kappa g,{}^{*a}\overline{\chi^{a}}\gamma^{\mu}\psi_{\mu} + \kappa^{2}g^{*}\bar{\psi}_{\mu}\sigma^{\mu\nu}\psi_{\nu L}) \\
&+ \frac{\kappa^{2}}{6}e^{-1}\phi\partial_{\mu}(e\bar{\psi}_{\nu}\gamma^{\nu}\psi^{\mu}) \\
&- \frac{1}{4}F_{\mu\nu}{}^{\alpha}F^{\mu\nu\alpha} - \frac{1}{2}\bar{\lambda}^{\alpha}\not{D}(\omega(e))\lambda^{\alpha} + \frac{1}{2}D^{\alpha}D^{\alpha} - \frac{\kappa}{2}\bar{\psi}_{\mu}\sigma^{\kappa\lambda}\gamma^{\mu}\lambda^{\alpha}F_{\kappa\lambda}{}^{\alpha} \\
&+ e^{-1}L^{quartic}
\end{aligned}
\qquad (A.30)
$$

where we here have grouped all quartic interactions:

$$e^{-1}L^{quartic} = \phi,_{ab}{}^{cd}(\overline{\chi}_a\chi^b)(\overline{\chi^c}\chi_d) - \kappa^2\phi,^a{}_b(\overline{\chi^a}\psi_\mu)(\overline{\psi}_\mu\chi^b)$$

$$- \frac{\kappa^2}{8}e^{-1}\varepsilon^{\mu\nu\rho\sigma}\overline{\psi}_\mu\gamma_\nu\psi_\rho(\phi,^a{}_b\overline{\chi^a}\gamma_\sigma\chi^b + \frac{1}{2}\overline{\lambda}^\alpha\gamma_5\gamma_\sigma\lambda^\alpha)$$

$$+ \frac{\kappa^2}{4}\overline{\lambda}^\alpha\gamma^\mu\sigma^{\nu\rho}\psi_\mu\overline{\psi}_\nu\gamma_\rho\lambda^\alpha$$

$$+ \frac{\kappa^4}{96}\phi((\overline{\psi}_\mu\gamma_\nu\psi_\sigma + 2\overline{\psi}_\nu\gamma_\mu\psi_\rho)\overline{\psi}^\mu\gamma^\nu\psi^\rho - 4(\overline{\psi}_\mu\gamma^\nu\psi_\nu)^2)$$

$$- \frac{\kappa^3}{6}(\overline{\psi}^\nu\gamma^\rho\psi_\rho - \frac{1}{4}e^{-1}\varepsilon^{\mu\nu\rho\sigma}\overline{\psi}_\mu\gamma_\rho\psi_\sigma)(\phi,^a\overline{\chi^a}\psi_\nu)$$

$$- \frac{\kappa^3}{6}(\overline{\psi}^\nu\gamma^\rho\psi_\rho + \frac{1}{4}e^{-1}\varepsilon^{\mu\nu\rho\sigma}\overline{\psi}_\mu\gamma_\rho\psi_\sigma)(\overline{\psi}_\nu\chi^a\phi,_a) \ . \qquad (A.31)$$

In arriving at Eqs. (A.30) and (A.31), we have used the following Fierz identities

$$\overline{\chi^a}\gamma^\mu\chi^b\psi_\mu\chi^c = \frac{1}{2}\overline{\psi}_\mu\gamma^\mu\chi_a\overline{\chi_b}\chi^c$$

$$\overline{\chi^a}\sigma^{\mu\nu}\sigma^{\rho\sigma}\psi_\nu K_{\mu\rho\sigma} = \frac{\kappa^2}{8}\overline{\chi^a}\psi_\nu(\overline{\psi}^\nu\gamma^\rho\psi_\rho - \frac{e^{-1}}{4}\varepsilon^{\mu\nu\rho\sigma}\overline{\psi}_\mu\gamma_\rho\psi_\sigma) \ . \qquad (A.32)$$

We now seperate from the Lagrangian all terms that include the auxiliary fields u, A_μ, h^a, and D^α. However, in analogy with the case of one chiral multiplet [C7], we find it more convenient to treat $u - \frac{\kappa}{2}d,_a h^a$ as an independent variable instead of u itself, where the function d is defined in Eq. 2.19. (It is related to the function J that appears in Ref. [C7] by $J = -\frac{\kappa^2}{2}d$). The auxiliary part of the Lagrangian now reads

$$e^{-1}L^{aux} = \frac{\kappa^2\phi}{9}\left|u - \frac{\kappa}{2}d,_a h^a\right|^2 - \frac{\kappa^2}{6}\phi, d^a{}_b h_a h^b$$

$$+ \frac{1}{2}[\kappa g(u - \frac{\kappa}{2}d,_a h^a) - \frac{\kappa^2}{9}\phi A_\mu A^\mu + g(\frac{\kappa^2}{2}d,_a + \frac{g,_a}{g})h^a + h\cdot c]$$

$$+ \frac{\kappa^2\phi}{6}\{[\frac{\kappa}{3}(u - \frac{\kappa}{2}d,_c h^c)(d,^{ab} - \frac{\kappa^2}{6}d,^a d,^b) + h^c(d,_c{}^{ab} - \frac{\kappa^2}{3}d,^a d,^b{}_c)]$$

$$\cdot(\bar{\chi}^a\chi_b) + h\cdot c\} - i\frac{\kappa^3\phi}{18}A^\mu(d,_a\hat{\mathcal{D}}_\mu z^a - d,^a\hat{\mathcal{D}}_\mu z_a$$

$$+ (d,^a{}_b - \frac{\kappa^2}{6}d,^a d,_b)\bar{\chi}^a\gamma_\mu\chi^b + \frac{9}{2\kappa^2\phi}\bar{\lambda}^\alpha\gamma_\mu\gamma_5\lambda^\alpha) \qquad \text{(A.33)}$$

The field equations for $u - \frac{\kappa}{2}d,_a h^a$, h^a, A_μ and D^α give respectively:

$$u - \frac{\kappa}{2}d,_a h^a = -\frac{9}{2\kappa\phi}g^* - \frac{\kappa}{2}(d,_{ab} - \frac{\kappa^2}{6}d,_a d,_b)\bar{\chi}_a\chi^b$$

$$h^a = \frac{3}{\kappa^2\phi}g^*(d^{-1})^a{}_b(\frac{\kappa^2}{2}d,^b + \frac{g,^{*b}}{g^*}) + ((d^{-1})^a{}_b d,^b{}_{ce} - \frac{\kappa^2}{3}\delta^a{}_c d,_e)\bar{\chi}_c\chi^e$$

$$A_\mu = -i\frac{\kappa}{4}[d,_a\hat{\mathcal{D}}_\mu z^a - d,^a\hat{\mathcal{D}}_\mu z_a + (d,^a b - \frac{\kappa^2}{6}d,^a d,_b)\bar{\chi}^a\gamma^\mu\chi^b + \frac{9}{2\kappa^2\phi}\bar{\lambda}^\alpha\gamma_\mu\gamma_5\lambda^\alpha]$$

$$D^\alpha = -\frac{g_\alpha}{2}\phi,_a(T^\alpha z)^a\ . \qquad \text{(A.34)}$$

Substituting back from Eq. (A.34) into the auxiliary part of the Lagrangian we obtain:

$$e^{-1}L^{aux} = -\frac{9}{4\phi}\left|g + \frac{\kappa^2\phi}{9}(d,^{ab} - \frac{\kappa^2}{6}d,^a d,^b)\bar{\chi}^a\chi_b\right|^2$$

$$+ \frac{6}{\kappa^2\phi}(d^{-1})^a{}_b[\frac{g}{2}(\frac{g,_a}{g} + \frac{\kappa^2}{2}d,_a) + \frac{\kappa^2\phi}{6}(d,^{ce}{}_a - \frac{\kappa^2}{3}d,^c{}_a d,^e)\bar{\chi}^c\chi_e]$$

$$\times[\frac{g^*}{2}(\frac{g^*,^b}{g^*} + \frac{\kappa^2}{2}d,^b) + \frac{\kappa^2\phi}{6}(d,^b{}_{c'e'} - \frac{\kappa^2}{3}d,^b{}_{c'}d,_{e'})\bar{\chi}_{c'}\chi^{e'}]$$

$$- \frac{\kappa^4\phi}{144}(d,_a \hat{\mathcal{D}}_\mu Z^a - d,^a \hat{\mathcal{D}}_\mu Z_a + (d,^a{}_b - \frac{\kappa^2}{6} d,^a d,_b)\overline{\chi}^a \gamma^\mu \chi^b + \frac{3}{2\kappa^2\phi}\bar{\lambda}^\alpha \gamma_\mu \gamma_5 \lambda^\alpha)^2$$

$$- \frac{g_\alpha^2}{8}(\phi,_a (T^\alpha Z)^a)^2 \tag{A.35}$$

Going back to Eq. (A30), the curvature scalar R appears with a factor $\phi/6$ implying, as has been mentioned earlier, the need for a Weyl-scaling to separate the physical graviton from the scalar fields. The required scaling is:

$$e_\mu{}^r \to \exp(\frac{\kappa^2 d}{12}) e_\mu{}^r \tag{A.36}$$

Then the connection and the curvature would scale according to:

$$\omega_\mu{}^{rs}(e) \to \omega_\mu{}^{rs}(e) + \frac{\kappa^2}{12}(e_\mu{}^r e^{\nu s} - e_\mu{}^s e^{\nu r})\partial_\nu d$$

$$e\frac{\phi}{6}R(\omega(e)) \to - \frac{e}{2\kappa^2}R(\omega(e)) - e\frac{\kappa^2}{48}(\partial_\mu d)(\partial^\mu d) \ . \tag{A.37}$$

and the bosonic part of the Lagrangian becomes as given in Eq. (2.22). Eq.(2.22) can be further simplified and expressed in terms of one function instead of two by defining:

$$\mathcal{G} = - \frac{\kappa^2}{2}d - \log\frac{\kappa^6}{4}|g|^2 \tag{A.38}$$

In terms of G the effective potential would read

$$e^{-1}V = - \frac{1}{\kappa 4}\exp(-\mathcal{G})((\mathcal{G}^{-1})^a{}_b \mathcal{G},_a \mathcal{G},^b + 3) + \frac{g_\alpha^2}{8\kappa^4}(\mathcal{G},_a (T^\alpha Z)^a))^2 \tag{A.39}$$

where we have used the fact that the superpotential g is gauge invariant implying for a gauge variation

$$\delta g(Z) = \varepsilon^{\alpha}(g,_a(T^{\alpha}Z)^a) = 0 \tag{A.40}$$

The fermionic part of the Lagrangian is more complicated. However, the required steps to arrive at the final result are well defined. Firstly, the gravitino field kinetic energy is mixed with those of the spin-1/2 fields due to the presence of the term

$$\phi,^a \overline{\chi}^a \sigma^{\mu\nu} D_\mu \psi_\nu + h \cdot c$$

in the Lagrangian. This implies that the gravitino field has to be redefined. Moreover, the spinors ψ_μ, λ^α, χ^a have to be rescaled to obtain the proper normalizations. The correct transformations are:

$$\chi^a = \exp(-\frac{\kappa^2 d}{24})(\frac{g^*}{g})^{\frac{1}{4}}(\chi^a)^{new}$$

$$\lambda_L^\alpha = \exp(-\frac{\kappa^2 d}{8})(\frac{g}{g^*})^{\frac{1}{4}}(\lambda_L{}^\alpha)^{new};\ \lambda_R^\alpha = \exp(-\frac{\kappa^2 d}{8})(\frac{g^*}{g})^{1/4}(\lambda_R^\alpha)^{new}$$

$$\psi_{\mu L} = (\frac{g}{g^*})^{\frac{1}{4}}[\exp(\frac{\kappa^2 d}{24})\psi_{\mu L}{}^{new} + \frac{\kappa}{6}\gamma_r(e_\mu^r)^{new}\ d,^a(\chi_a)^{new}]$$

$$\psi_{\mu R} = (\frac{g^*}{g})^{\frac{1}{4}}[\exp(\frac{\kappa^2 d}{24})\psi_{\mu R}{}^{new} + \frac{\kappa}{6}\gamma_r(e_\mu^r)^{new} d,_a(\chi^a)^{new}] \tag{A.41}$$

where $(e_\mu^r)^{new}$ is the new vierbein of Eq. (A.36). Substituting Eqs. (A.41) into (A.30) and (A.35) we obtain after a lengthy algebra (and dropping the index "new"):

$$
\begin{aligned}
L_F = &- \frac{1}{2}\epsilon^{\mu\nu\rho\sigma}\bar{\psi}_\mu\gamma_5\gamma_\nu D_\rho(\omega(e,\psi))\psi_\sigma + \frac{e}{\kappa^2}\mathcal{G}^{,a}{}_{b}\overline{\chi^a}\not{\mathcal{D}}\chi^b - \frac{e}{2}\bar{\lambda}^\alpha\not{D}\lambda^\alpha \\
&- e\frac{\kappa}{2}\bar{\psi}_\mu\sigma^{\kappa\lambda}\gamma^\mu\lambda^\alpha F_{\kappa\lambda}{}^\alpha + \frac{1}{8}\epsilon^{\mu\nu\rho\sigma}\bar{\psi}_\mu\gamma_\nu\psi_\rho(\mathcal{G}_{,a}\mathcal{D}_\sigma z^a - \mathcal{G}^{,a}\mathcal{D}_\sigma z_a) \\
&- \frac{e}{\kappa}\mathcal{G}^{,a}{}_{b}(\mathcal{D}_\nu z^b\overline{\chi^a}\gamma^\mu\gamma^\nu\psi_\mu + \bar{\psi}_\mu\gamma^\nu\gamma^\mu\chi^b\mathcal{D}_\nu z_a) \\
&+ \frac{e}{\kappa^2}\overline{\chi^a}\gamma^\mu\chi^b((\mathcal{G}^{,a}{}_{bc} - \frac{\kappa^2}{4}\mathcal{G}^{,a}{}_{b}\mathcal{G}_{,c})\mathcal{D}_\mu z^c - (\mathcal{G}^{,ac}{}_{b} - \frac{\kappa^2}{4}\mathcal{G}^{,a}{}_{b}\mathcal{G}^{,c})\mathcal{D}_\mu z_c) \\
&+ \frac{e}{8}\bar{\lambda}^\alpha\gamma^\mu\gamma_5\lambda^\alpha(\mathcal{G}_{,a}\mathcal{D}_\mu z^a - \mathcal{G}^{,a}\mathcal{D}_\mu z_a) \\
&+ \frac{e}{\kappa^3}\exp(-\frac{\mathcal{G}}{2})[(\mathcal{G}_{,ab} - \mathcal{G}_{,a}\mathcal{G}_{,b} - (\mathcal{G}^{-1})^c{}_d\mathcal{G}_{,c}\mathcal{G}^{,d}{}_{ab})\bar{\chi}_a\chi^b \\
&- \kappa\mathcal{G}_{,a}\bar{\psi}_\mu\gamma^\mu\chi^a + \kappa^2\bar{\psi}_\mu\sigma^{\mu\nu}\psi_{\nu R} + h\cdot c] \\
&+ ie\frac{g_\alpha}{4\kappa}(\mathcal{G}_{,a}(T^\alpha z)^a\bar{\psi}_\mu\gamma_5\gamma^\mu\lambda^\alpha + ei\frac{g_\alpha}{\kappa^2}\mathcal{G}^{,a}{}_{b}((T^\alpha z)_a\bar{\lambda}^\alpha\chi^b - \overline{\chi^a}\lambda^\alpha(T^\alpha z)^b) \\
&- \frac{e}{\kappa^2}(\mathcal{G}_{,ab}{}^{cd} - (\mathcal{G}^{-1})^e{}_f\mathcal{G}_{,ab}{}^{e}\mathcal{G}^{,cd}{}_{f} + \frac{1}{2}\mathcal{G}^{,c}{}_{a}\mathcal{G}^{,d}{}_{b})\overline{\chi_a}\chi^b\overline{\chi^c}\chi_d \\
&+ e\frac{\kappa^2}{4}\bar{\lambda}^\alpha\gamma^\mu\sigma^{\nu\rho}\psi_\mu\bar{\psi}_\nu\gamma_\rho\lambda^\alpha + \frac{e}{8}\mathcal{G}^{,a}{}_{b}\overline{\chi^a}\gamma^\mu\chi^b\bar{\lambda}^\alpha\gamma_\mu\gamma_5\lambda^\alpha \\
&+ 3e\frac{\kappa^2}{64}\bar{\lambda}^\alpha\gamma^\mu\gamma_5\lambda^\alpha\bar{\lambda}^\beta\gamma_\mu\gamma_5\lambda^\beta + \frac{1}{4}\mathcal{G}^{,a}{}_{b}\overline{\chi^a}\gamma_\sigma\chi^b(\epsilon^{\mu\nu\rho\sigma}\bar{\psi}_\mu\gamma_\nu\psi_\rho - e\bar{\psi}_\mu\gamma_5\gamma^\sigma\psi^\mu) . \qquad (A.42)
\end{aligned}
$$

The Fermionic Lagrangian (A.42) depends only on the function . In arriving at (A.42) we grouped all terms of the same form together, first expressing them as a function of g and d, and finally simplifying them into a function of $\mathcal{G}$.

The resultant Lagrangian is invariant under supersymmetry transformations only on the smass-shell. To obtain the supersymmetry transformations we substitute the auxiliary fields from Eq. (A.34) into the old transformations and substitute Eq. (A.41). The new transformations are:

$$\delta_s e_\mu{}^r = \kappa\bar{\varepsilon}\gamma^r\psi_\mu$$

$$\delta_s\psi_{\mu L} = 2\kappa^{-1}D_\mu(e,\psi_\nu)\varepsilon_L + \kappa^{-2}\exp(-\tfrac{\mathcal{G}}{2})\gamma_\mu\varepsilon_R + \tfrac{1}{2}(\mathcal{G},_a\bar{\varepsilon}\chi^a - \mathcal{G},^a\overline{\chi^a}\varepsilon)\psi_{\mu L}$$

$$+ \kappa^{-1}(\sigma_{\mu\nu}\varepsilon_L)\mathcal{G},^a{}_b\overline{\chi^a}\gamma^\nu\chi^b - \tfrac{\kappa}{8}(\delta_\mu{}^\nu + \gamma^\nu\gamma_\mu)\varepsilon_L\bar{\lambda}^\alpha\gamma_\nu\gamma_5\gamma^\alpha$$

$$- \frac{\kappa^{-1}}{2}(\mathcal{G},_a\mathcal{D}_\mu z^a - \mathcal{G},^a\mathcal{D}_\mu z_a)\varepsilon_L$$

$$\delta_s V_\mu{}^\alpha = -\bar{\varepsilon}\gamma_\mu\lambda^\alpha$$

$$\delta_s\lambda^\alpha = -\tfrac{1}{2}(\mathcal{G},_a\bar{\varepsilon}\chi^a - \mathcal{G},^a\overline{\chi^a}\varepsilon)\lambda^\alpha - \sigma^{\mu\nu}\varepsilon\hat{F}_{\mu\nu}{}^\alpha - \frac{ig_\alpha}{2\kappa^2}(\gamma_5\varepsilon)(\mathcal{G},_a(T^\alpha z)^a)$$

$$\delta_s z^a = 2\bar{\varepsilon}\chi^a$$

$$\delta_s\chi^a = \gamma^\mu\varepsilon_R\hat{\mathcal{D}}_\mu z^a - \tfrac{1}{2}(\mathcal{G},_b\bar{\varepsilon}\chi^b - \mathcal{G},^b\overline{\chi^b}\varepsilon)\chi^a$$

$$+ (\mathcal{G}^{-1})^a{}_b\mathcal{G},^b{}_{cd}\overline{\chi_c}\chi^d\varepsilon_L - \kappa^{-1}(\mathcal{G}^{-1})^a{}_b\mathcal{G},^b\exp(-\tfrac{\mathcal{G}}{2})\varepsilon_L \quad . \qquad \text{(A.43)}$$

From the Lagrangian in (2.21) and (A.42) it is possible to read the mass matrices of the physical fields

$$\psi_\mu,\ \lambda^\alpha,\ \chi^a,\ z^a,\ \text{and}\ V_\mu{}^\alpha . \qquad \text{(A.44)}$$

The gauge bosons $V_\mu{}^\alpha$ that correspond to broken generators of the gauge group acquire their masses by the usual Higgs mechanism, while those corresponding to unbroken generators remain massless. The mass formula for the massive gauge bosons are the same as in standard gauge theories.

The mass formula for the complex fields Z^a must be given in terms of the real fields A^a and B^a, because their masses will split when supersymmetry is broken. We can expand V in terms of the complex fields:

$$V(Z^a, Z_b) = \frac{1}{2}(V,_{ab})_o Z^a Z^b + \frac{1}{2}(V,^{ab})_o Z_a Z_b + (V,^a{}_b)_o Z_a Z^b + \ldots \tag{A.45}$$

Alternatively, we can write

$$V(Z^a, Z_b) = \frac{1}{2}((M^2{}_{ab})^A A^a A^b + (M^2)^B{}_{ab} B^a B^b) + \ldots$$

Thus:

$$(M^2)^A{}_{ab} = (V,^a{}_b + V,^b{}_a + 2V,_{ab})_o$$

$$(M^2)^B{}_{ab} = (V,^a{}_b + V,^b{}_a - 2V,_{ab})_o \; . \tag{A.46}$$

It is useful, and considerably much simpler, to express the above equations as a function of $\mathcal{G}$ for the case of a flat Kähler manifold:

$$V,_{ab} = \frac{2e}{\kappa^4}\exp(-\mathcal{G})\left[\frac{1}{2}(\mathcal{G},_{ab} - \mathcal{G},_a\mathcal{G},_b) + \frac{1}{\kappa^2}(\mathcal{G},_{abc} - 3\mathcal{G},_{(a}\,\mathcal{G},_{bc)} + \mathcal{G},_a\mathcal{G},_b\mathcal{G},_c)\mathcal{G},^c\right] + \frac{g_\alpha^2}{32} Z_c Z_d (T^\alpha)^c{}_a (T^\alpha)^d{}_b$$

$$V,^a{}_b = \frac{2e}{\kappa^4}\exp(-\mathcal{G})[-\frac{1}{2}\mathcal{G},^a\mathcal{G},_b + \frac{1}{\kappa^2}(\mathcal{G},_{bc} - \mathcal{G},_b\mathcal{G},_c)(\mathcal{G},^{ac} - \mathcal{G},^a\mathcal{G},^c)$$

$$-\frac{\kappa^2}{2}\delta^a{}_b + \frac{1}{2}\delta^a{}_b\mathcal{G},_c\mathcal{G},^c] + \frac{g_\alpha^2}{32}z_c\bar{z}^d(T^\alpha)^a{}_d \; . \tag{A.47}$$

The mass terms for the fields ψ_μ and χ^a are (for a flat Kähler manifold)

$$L^{mass} = \frac{e}{\kappa^3}\exp(-\frac{\mathcal{G}}{2})[(\mathcal{G},_{ab} - \mathcal{G},_a\mathcal{G},_b)\overline{\chi_a}\chi^b - \kappa\mathcal{G},_a\bar{\psi}_\mu\gamma^\mu\chi^a + \kappa^2\bar{\psi}_\mu\sigma^{\mu\nu}\psi_{\nu R} + h\cdot c]$$

$$-ei\frac{g_\alpha}{8}\kappa(z_a(T^\alpha z)^a)\bar{\psi}_\mu\gamma_5\gamma^\mu\lambda^\alpha - ei\frac{g_\alpha}{2}((T^\alpha z)_a\bar{\lambda}^\alpha\chi^a - \overline{\chi^a}\lambda^\alpha(T^\alpha z)^a) \tag{A.48}$$

If supersymmetry is broken, the gravitino field will be described by:

$$\psi'_{\mu R} = \psi_{\mu R} - \frac{\kappa^{-1}}{3}\gamma_\mu(\mathcal{G},_a\chi^a + i\frac{g_\alpha\kappa^3}{4}\exp(-\frac{\mathcal{G}}{2})(z_a(T^\alpha z)^a)\lambda_L^\alpha) \; . \tag{A.49}$$

After substituting the last equation into (A.48) the mass part of the Lagrangian becomes

$$L^{mass} = \frac{e}{\kappa}\exp(-\frac{\mathcal{G}}{2})(\bar{\psi}'_\mu\sigma^{\mu\nu}\psi'_{\nu L}) - \frac{e}{2}(m_{ab}\overline{\chi_a}\chi^b) - em^\alpha{}_a\bar{\lambda}^\alpha\chi^a$$

$$-\frac{e}{2}m^{\alpha\beta}\bar{\lambda}_L{}^\alpha\lambda_L{}^\beta + h\cdot c \tag{A.50}$$

where

$$m_{ab} = \frac{2}{\kappa^3}\exp(-\frac{\mathcal{G}}{2})(\mathcal{G},_{ab} - \frac{1}{3}\mathcal{G},_a\mathcal{G},_b)$$

$$m^\alpha{}_a = i\frac{g_\alpha}{2}z_b(T^\alpha)^b{}_a$$

$$m^{\alpha\beta} = -\exp(-\frac{\mathcal{G}}{2})\frac{g_\alpha^2\kappa^3}{24}(z_a(T^\alpha z)^a)(z_b(T^\beta z)^b) \tag{A.51}$$

By using the minimization condition

$$V,_a = 0 \tag{A.52}$$

and the vanishing of the cosmological constant

$$V = 0$$

which are equivalent to

$$\mathcal{G},_{ab}\mathcal{G},^b = \frac{\kappa^2}{2}\mathcal{G},_a$$

$$(Z_a(T^\alpha Z)^a) = 0$$

$$\mathcal{G},_a\mathcal{G},^a = \frac{3}{2}\kappa^2 \tag{A.53}$$

one can easily prove that [C9]

$$\text{supertrace } M^2 = \sum_{J=0}^{3/2}(-1)^{2J}(2J+1)m_J^2 = 2(N-1)m_{3/2}^2 \tag{A.54}$$

where N is the number of chiral multiplets. This can be seen by noting that the gauge particle contributions are the same as in global supersymmetry while the rest (obtained by letting $g_\alpha = 0$) are given by

$$\begin{aligned} \mathrm{tr}(M^2)^A + \mathrm{tr}(M^2)^B - 2\,\mathrm{tr}\, m_{ab}^2 - 4m_{3/2}^2 = 4V,^a{}_a - \frac{8}{\kappa^6}\exp(-\mathcal{G})(\mathcal{G},_{ab} - \tfrac{1}{3}\mathcal{G},_a\mathcal{G},_b) \\ \cdot\, (\mathcal{G},^{ab} - \tfrac{1}{3}\mathcal{G},^a\mathcal{G},^b) - \frac{4}{\kappa^2}\exp(-\mathcal{G}) \end{aligned} \tag{A.55}$$

This is equal to, after substituting in Eqs. (A.47) and (A.53),

$$2\kappa^{-2}\ (N - 1)\ \exp\ (-\mathcal{G})\ . \tag{A.56}$$

Finally, we note that the existence of different sets of auxiliary fields for N = 1 supergravity, poses the question whether the results we obtained remain valid for the other sets.

Kugo and Uehara [C11] have introduced a method based on superconformal tensor calculus [C15] that can accomodate the different sets of auxiliary fields. It was later proved [C12] that all interactions constructed in terms of the other known sets of auxiliary fields are particular examples of the interactions constructed here with the minimal set.

REFERENCES

A. GLOBAL SUPERSYMMETRY

[A1] J. Wess and B. Zumino, Nucl. Phys. B70 39 (1974).

[A2] A. Salam and J. Strathdee, Phy. Rev. D11, 1521 (1975).

[A3] For a review of supersymmetry see A. Salam and J. Strathdee, Forts. Phys. 26 (1978) 57.
P. Fayet and S. Ferrara, Phys. Reports 32C (1977) 249.

[A4] M.T. Grisaru, W. Siegel and M. Rocek, Nucl. Phys. B159 (1979) 420.

B. SUSY MODELS

[B1] P. Fayet, Phys. Lett 69B (1977) 489, 70B (1977) 461.

[B2] E. Witten, Nucl. Phys. B 177 (1981), B185, 513 (1981).

[B3] M. Dine, W. Fischler and M. Srednicki, Phys. Lett. 104B 199 (1981) and Nucl. Phys. B192, 353 (1981).

[B4] S. Dimopoulos and H. Georgi, Nucl. Phys. B193, 150 (1981).

[B5] N. Sakai, Z. Phys. C 11, 153 (1981).

[B6] D.V. Nanopoulos and K. Tamvakis, Phys. Lett. 110B (1982) 449.

[B7] R.K. Kaul, Phys. Lett. 109B, 19 (1982). R.K. Kaul and P. Majumdar, Nucl. Phys. B199 (1982) 36.

[B8] L. Alvarez-Gaume, M. Claudson and M. Wise, Nucl. B207, 96 (1982).

[B9] C. Nappi and B. Ovrut, Phys. Lett. 113B, 175 (1982).

[B10] L. Ibanez and G.G. Ross, Phys. Lett. 110B (1982) 215.

[B11] J. Ellis, L. Ibanez and G.G. Ross, Phys. Lett. 113B, 283 (1982), Nucl. Phys. B221 (1983) 29.

[B12] S. Dimopoulos and S. Raby, Nucl. Phys. B192 (1981) 353.

[B13] J. Polchinski and L. Susskind, Phys. Rev. D26 (1982) 3661.

[B14] A. Masiero, D.V. Nanopoulos, K. Tamvakis and T. Yanagida, Phys. Lett. 115B (1982) 298.

[B15] S. Dimopoulos and F. Wilczek, Santa Barbara preprint (1981).

[B16] L. Hall, I. Hinchliffe and R. Cahn, 109B (1982) 426.

[B17] G. Ferrar and S. Weinberg, Phys. Rev. D27 (1983) 2732.

[B18] P. Frampton and Jihn E. Kim, Univ. N. Carolina preprint IFP-178-UNC.

[B19] M.B. Einhorn and D.R.T. Jones, Univ. of Michigan preprint UMHE 81-55 (1981).

[B20] W. Marciano and J. Senjanovic, Brookhaven preprint (1981).

[B21] S. Dimopoulos, S. Raby and F. Wilczek, Phys. Lett. 112B, 113 (1982).

[B22] J. Ellis, D.V. Nanopoulis and S. Rudaz, Nucl. Phys. B202 (1982) 43.

[B23] C.S. Aulakh and R. Mohapatra, Phys. Lett. 121B (1983) 147; CCNY-HEP 82/4.

[B24] K. Inoue, A. Kakuto, H. Komatsu and S. Takeshita, Prog. Theor. Phys. 68 927 (1982).

[B25] M. Dine and W. Fischler, Phys. Lett. 110B (1982) 227.

[B26] B. Grinstein, Nucl. Phys. B206 (1982) 387.

[B27] J. Leon, M. Quiros and M. Ramon Medrano, Nucl. Phys. B222 (1983) 104.

C. SUPERGRAVITY TENSOR CALCULUS AND MATTER COUPLINGS

[C1] D.Z. Freedman, P. van Nieuwenbuizen and S. Ferrara, Phys. Rev. 1013, 3214 (1976); S. Deser and B. Zumino, Phys. Lett. 62B, 335 (1976).

[C2] K.S. Stelle and P.C. West, Nucl. Phys. B145, 175 (1978).

[C3] P. van Nieuwenhuizen and S. Ferrara, Phys. Lett. 76 (1978) 404.

[C4] For a review of Supergravity, see P. van Nieuwenhuizen, Phys. Reports 68 (4), 1981.

[C5] S. Deser and B. Zumino, Phys. Rev. Lett. 38 (1977) 1433.

[C6] J. Polony, University of Budapest Report No. KFKI-1977-93, 1977 (unpublished).

[C7] E. Cremmer, B. Julia, J. Scherk, S. Ferrara, L. Girardello and P. van Niewwenhuizen, Nucl. Phys. B147, 105 (1979).

[C8] A.H. Chamseddine, R. Arnowitt and P. Nath, Phys. Rev. Lett. 49, 970 (1982).

[C9] E. Cremmer, S. Ferrara, L. Girardello and A. van Proeyen, Phys. Lett. 116, B231 (1982); Nucl. Phys. B212 (1983) 413.

[C10] E. Witten and J. Bagger, Phys. Lett. 115B (1982) 202. J. Bagger, Nucl. Phys. B211 (1983) 302.

[C11] T. Kugo and S. Uehara, Nucl. Phys. B222 (1983) 125.

[C12] S. Ferrara, L. Girardello, T. Kugo and A. Van Proeyen, CERN TH 3523 (1983).

[C13] C.S. Aulakh, M. Kaku and R.N. Mohapatra, Preprint CUNY (1983).

[C14] B. Zumino, Phys. Lett. 87B, 203 (1979).

[C15] M. Kaku, P.K. Townsend and P. van Nieuwenhuizen, Phys. Rev. D17 (1978) 3179.

B. deWit, in "Supergravity 82", eds. S. Ferrara, J. Taylor and P. van Nieuwenhuizen, World Scientific Publishing Co.

A. Van Proeyen, TH 3579-CERN.

D. SU(2)XU(1) BREAKING AND SUPERGRAVITY MODELS

[D1] A.H. Chamseddine, R. Arnowitt and P. Nath, Phys. Rev. Lett. 49 (1982) 970.

[D2] P. Nath, R. Arnowitt and A.H. Chamseddine, Phys. Lett. 121B (1983) 33.

[D3] R. Arnowitt, A.H. Chamseddine and P. Nath, Phys. Lett. 120B, 145 (1983).

[D4] R. Barbieri, S. Ferrara and C.A. Savoy, Phys. Lett. 119B (1982) 343.

[D5] L.E. Ibanez, Phys. Lett. 118B (1982) 73 and Madrid preprint FTUAM/82-8.

[D6] J. Ellis, D.V. Nanopoulos and K. Tamvakis, Phys. Lett. 121B (1983) 123.

[D7] H.P. Nilles, M. Srednicki and D. Wyler, Phys. Lett. 124B (1983) 337 and 120B (1982) 346.

[D8] A.B. Lahanas, Phys. Lett. 124B (1983) 341.

[D9] N. Ohta, Prog. Theo. Phys. 70 (1983) 542.

[D10] P. Nath, A.H. Chamseddine and R. Arnowitt, Nucl. Phys. B227 (1983) 121).

[D11] L. Hall, J. Lykken and S. Weinberg, Phys. Rev. D27 (1983) 2359.

[D12] E. Cremmer, P. Fayet and L. Girardello, Phys. Lett. 112B, 41 (1983).

[D13] S.K. Soni and H.A. Weldon, Univ. of Pennsylvania Preprint 2/2/83.

[D14] C.S. Aulakh, CCNY-HEP-83/2.

[D15] B. Ovrut and S. Raby, Phys. Lett. 125B, (1983) 270.

[D16] J. Ellis, J. Hagelin, D. Nanopoulos and K. Tamvakis, Phys. Lett. 125B (1983) 275.

[D17] L. Alvarez-Gaume, J. Polchinski and M. Wise, Nucl. Phys. B221 (1983) 495. J. Polchinski, HUTP-83/A036.

[D18] L.E. Ibanez and C. Lopez, Phys. Lett. 126B, (1983) 94.

[D19] R. Arnowitt, A.H. Chamseddine and P. Nath, Proc. of Workshop on Problems in Unification and Supergrvity, LaJolla Institute, Jan. 1983.

[D20] S. Ferrara, D.V. Nanopoulos and C.A. Savoy, Phys. Lett. 123B (1983) 495.

[D21] J. Frere, M. Jones and S. Raby, Nucl. Phys. B222 (1983) 11.

[D22] S. Weinberg, Phys. Rev. Lett. 48 (1982) 1176.

[D23] B. Ovrut and J. Wess, Phys. Lett. 112B (1982) 347.

[D24] S. Coleman and F. DeLuccia, Phys. Rev. D21, 3305 (1980). L.F. Abbot and S. Deser, Nucl. Phys. B195, 76 (1982).

[D25] R. Barbieri, S. Ferrara, D.V. Nanopoulos and K.S. Stelle, Phys. Lett. 113B, 219 (1982).

[D26] C. Kounnas, A.B. Lahanas, D.V. Nanopoulos and M. Quiros, TH 3651-CERN (1983).

[D27] The statement in the second paper of [D7] regarding the existence of a saddle point in [D2] is incorrect.

E. PHENOMENOLOGICAL IMPLICATIONS OF SUPERGRAVITY UNIFIED MODELS

[E1] A.H. Chamseddine, R. Arnowitt and P. Nath, Phys. Rev. Lett. 49 (1982) 970.

[E2] R. Arnowitt, P. Nath and A.H. Chamseddine, Proc. of the International Symposium Gauge Theory and Gravitation, Tezukayama University, Nara, Japan August, 1982.

[E3] P. Nath, A.H. Chamseddine and R. Arnowitt, Proc. of APS Meeting at the University of Maryland, College Park, Maryland, October 1982.

[E4] S. Weinberg, Phy. Rev. Lett. 50 (1983) 387.

[E5] R. Arnowitt, A.H. Chamseddine and P. Nath, Phy. Rev. Lett. 50 (1983) 232.

[E6] P. Nath, A.H. Chamseddine and R. Arnowitt, Proc. of Orbis Scientiae, Miami, Florida, 1983.

[E7] M.-M. Frere and G.L. Kane, Nucl. Phys. B223, 331 (1983). G.L. Kane, in Proc. of the Fourth Workshop on Grand Unification (1983).

[E8] J. Ellis, J.S. Hagelin, D.V. Nanopoulos and M. Srednicki, Phys. Lett. 127B, 233 (1983).

[E9] A.H. Chamseddine, P. Nath and R. Arnowitt, Phys. Lett. 129B, 445 (1983).

[E10] J. Ellis and D.V. Nanopoulos, Phys. Lett. 116B (1982) 133.

[E11] N.S. Craigie, K. Hidaka and P. Ratcliffe, Phys. Lett. 129B (1983) 310.

[E12] P. Nath, R. Arnowitt and A.H. Chamseddine, Model Independent Analyses of Low Energy Supergravity Unified Theory, HUPT-83/A007/ NUB #2588.

[E13] P. Fayet, Phys. Lett. 117B, 460 (1983).

[E14] J. Ellis and J.S. Hagelin, Phys. Lett. 122B, 303 (1983).

[E15] R. Weinstein (Private Communication).

[E16] S. Weinberg, Phys. Rev. Lett. 48, 1303 (1982) J. Ellis, A.D. Linde and D.V. Nanopoulos, Phys. Lett. 118B (1982) 59.
L.M. Krauss, HUTP-83/A009.

[E17] H. Goldberg, Phys. Rev. Lett. 50 (1983) 1419.

[E18] R. Barbieri, L. Girardello and A. Marsiero, Phys. Lett. 127B (1983) 429.

[E19] B.W. Lee and S. Weinberg, Phys. Rev. Lett. 39 (1977) 165.

[E20] T. Veltman, Nucl. Phys. B123 (1977) 89.

[E21] The result of Eq. (5.15) is based on data extracted from the deep inelastic neutrino scattering on isoscalar targets and is an average of several analyses see W.J. Marciano and Z. Parsa in [F5]; J.E. Kim, P. Langacker, M. Lavine and H.H. Williams, Rev. of Mod. Phys. 53, NO.2 (1981) 211; M. Jonker et. al., Phys. Lett. 99B (1981) 265.

[E22] E. Eliasson "Radiative Corrections to the ρ Parameter In Supergravity Unified Theory", Northeastern University Preprint NUB#2621.

[E23] R. Barbieri and L. Maiani, Nucl. Phys. B224 (1983) 32.

[E24] D.A. Dicus, S. Nandi, W.W. Repko and X. Tata, Phys. Rev. Lett. 51 (1983) 1030.

[E25] R.M. Barnett, K. Lackner, H.E. Haber, Phys. Rev. Lett. 51 (1983) 176 and Phys. Lett. 126B, 64 (1983).

[E26] N. Cabibbo, L. Maiani and S. Petrarca, University of Rome Preprint n. 355 (1983); R. Barbieri, N. Cabibbo, L. Maiani and S. Petrarca, Phys. Lett. 127B, 458 (1983).

[E27] M. Glück and E. Reya, Phys. Rev. Lett. 51 (1983) 867.

[E28] G.L. Kane and W. Rolnick, UMTH83-14.

[E29] J. Polchinski and M. Wise, Phys. Lett. 125B, 393 (1983); R. Mohapatra, S. Ouvry and G. Senjanovic, Phys. Lett. 126B, 329 (1983), F. del Aguila, M.B. Gavela, J.A. Grifols and A. Mendez, Phys. Lett. 126B, 71 (1983), F. del Aguila, J.A. Grifols, A. Mendez, D. Nanopoulos and M. Srednicki, Phys. Lett. 129B, 77 (1983).

[E30] A.D. Linde, Talk at the Shelter Island Conference II, 1-3 June 1983. K.A. Olive, in Proc. of the 3rd Moriond Astrophysics Meeting, LaPlagne, France, 1983, Eds.J. Audouze and J. Tran Thanh Van.

F. GENERAL REFERENCES

[F1] S. Weinberg, Phys. Rev. Lett. 19 (1967) 1264.
A. Salam, Proc. 8th Nobel Symposium, Aspenäsgården, 1968 (Almqvist and Wiksell, Stockholm, 1968), p. 367.

[F2] C. Arnison et. al. Phys. Lett. 122B (1983) 103 G. Banner et. al. Phys. Lett. 122B (1983) 476.

[F3] C. Arnison et. al., Phys. Lett. 129B, (1983) 273.

[F4] S. Weinberg, in General Relativity - An Einstein Centenany Survey, edited by S.W. Hawking and W. Israel (Cambridge University Press, Cambridge, England, 1979), Chap. 16.

[F5] Proc. of 1982 DPF-Summer Study on Elementary Particle Physics and Future Facilities, June 29-July 16, 1982, Snowmass Colorado, edited by Rene Donaldson, Richard Dustafson and Frank Paige.

[F6] R.M. Godbole, S. Pakvasa and D.P. Roy, Phys. Lett. 50 (1983) 1539. V. Barger, A.D. Martin and R.J.N. Phillips, Phys. Lett. 125B, 339 (1983) and MAD/PH/100 (1983).

[F7] For a review of the current status see lectures by J.C. Pati at this meeting "Summer Workshop on Particle Physics, Trieste, June-July 1983" and also in Proc. of the ICOMAN Conference, Frascati, January 1983.

Figure Captions

Fig. (1): $e^+e^- \rightarrow \gamma\tilde{\gamma}\tilde{\gamma}$ process [Fig. (1a)] and competing background $e^+e^- \rightarrow \gamma\nu\bar{\nu}$ [Fig. (1b)].

Fig. (2): $e^+e^- \rightarrow e^+e^-\tilde{\gamma}\tilde{\gamma}$ process [Fig. (2a)] and competing background $e^+e^- \rightarrow e^+e^-\nu\bar{\nu}$.

Fig. (3): Radiative corrections to the guagino masses arising from Eq. (5.10) [Fig. (3a)] and from Eq. (5.11) [Fig. (3b)].

Fig. (4): Source of significant direct gaugino masses due to the exchange of heavy fields of the GUT sector.

Fig. (5): The ρ parameter as a function of the top quark mass in Supergravity unified theory.

Fig. (6): Decay modes of the Winos and Zinos. In addition to the squark ($\tilde{q}$) intermediate states, the decays can proceed also through sleptons when the final states contain a photino ($\tilde{g}$ = gluino, l = lepton).

Fig. (7): Decays with Unidentified Fermionic Objects (UFO), the photinos, with one and two jets.

Fig. (8): Lepton-jet decays with unbalanced $p_\perp$.

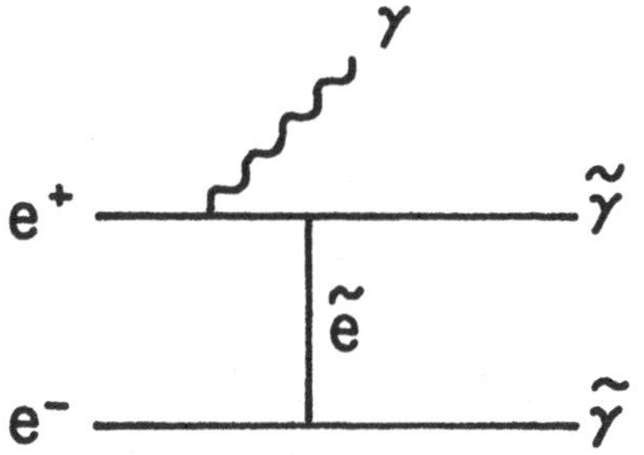

Fig. (1a.)

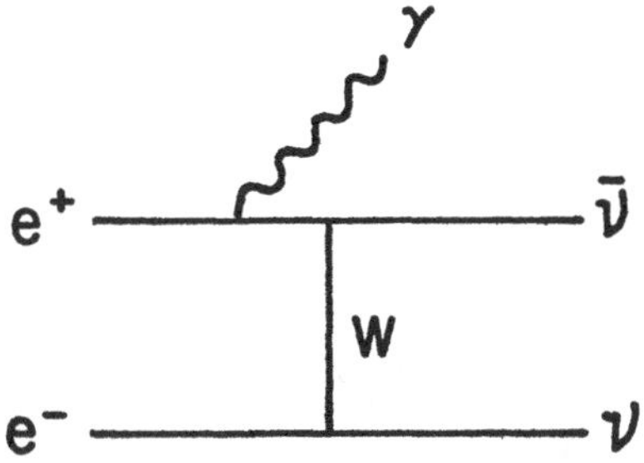

Fig. (1b.)

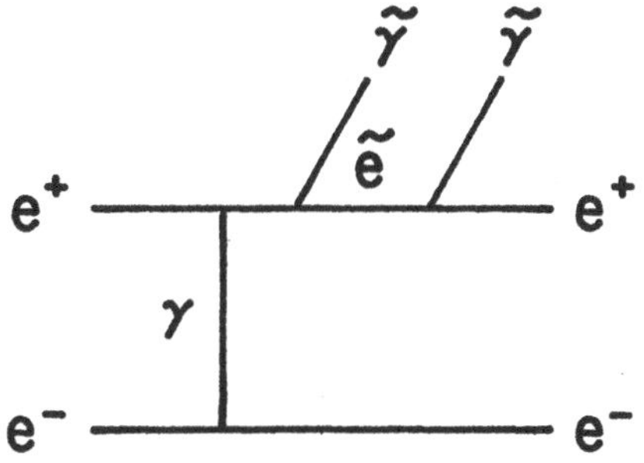

Fig. (2a.)

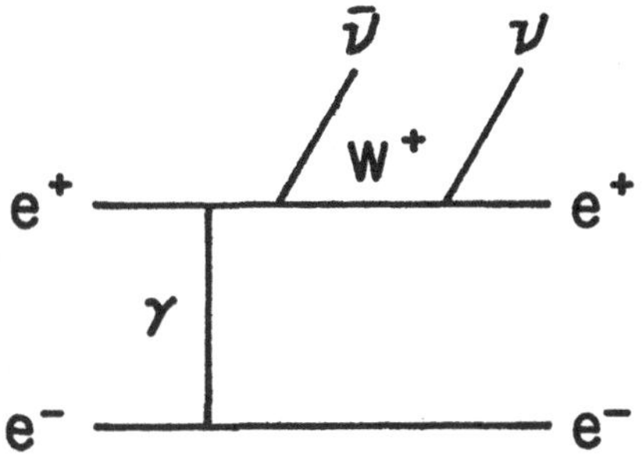

Fig. (2b.)

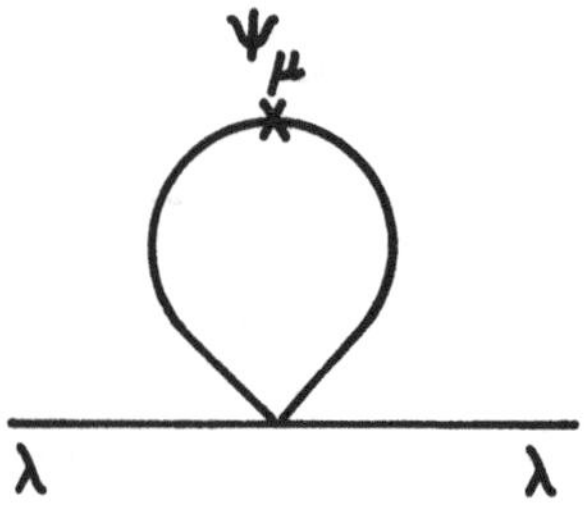

Fig (3a)

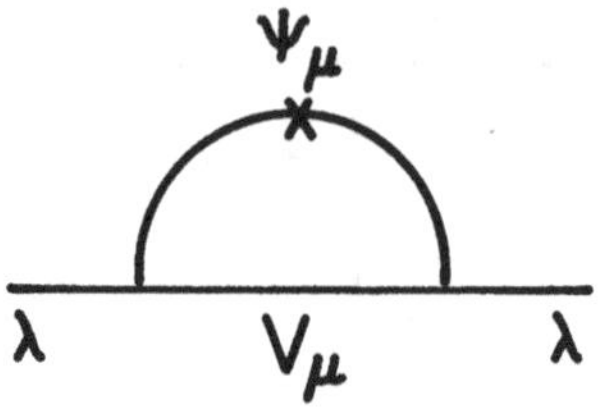

Fig (3b)

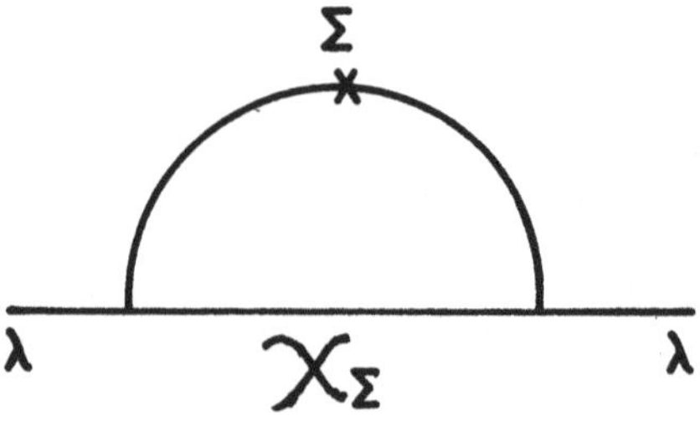

Fig. (4)

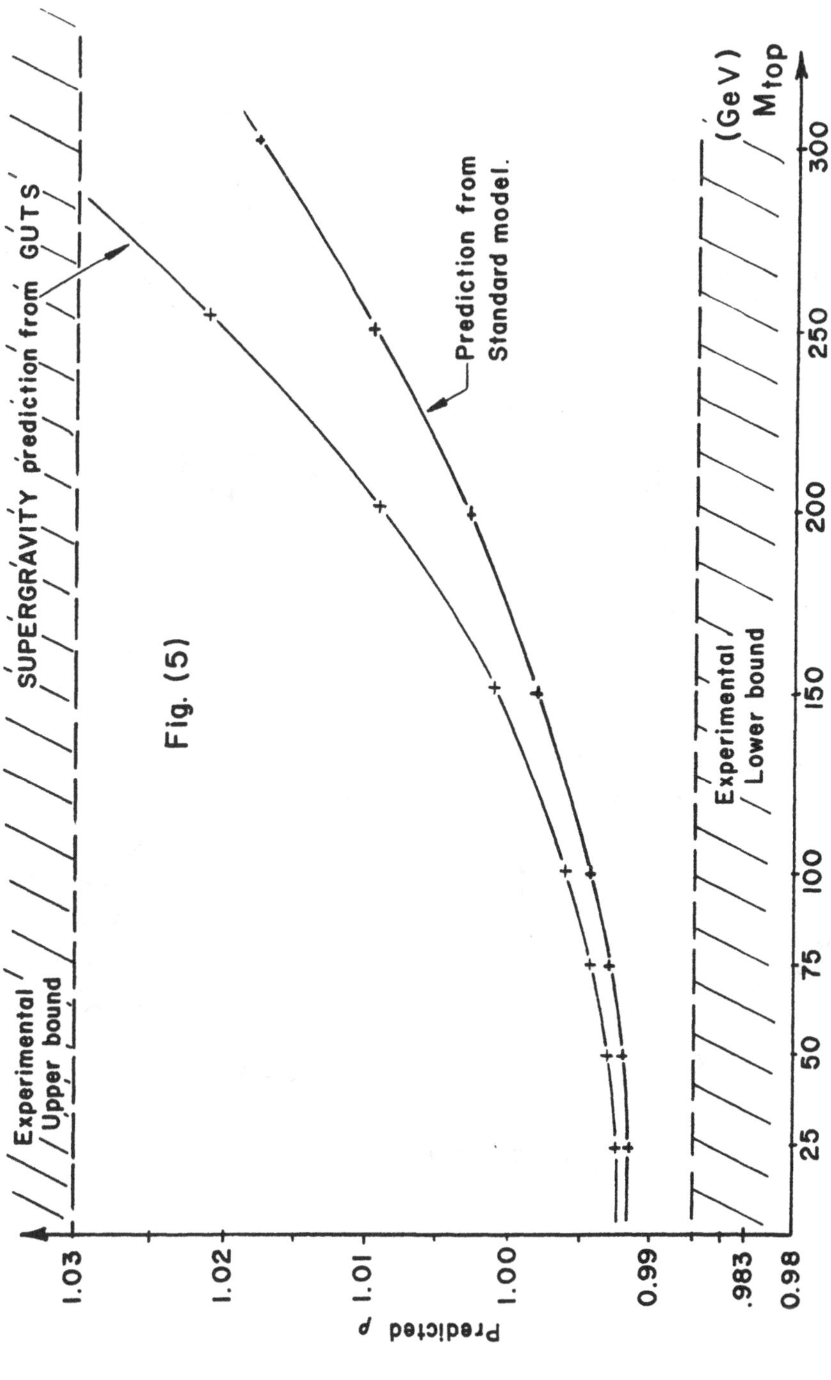
Experimental
Upper bound
SUPERGRAVITY prediction from GUTS
Prediction from
Standard model.
Experimental
Lower bound
(GeV)
Mtop
25
50
75
100
150
200
250
300
1.03
1.02
1.01
1.00
0.99
.983
0.98
Predicted ρ

Fig. (5)

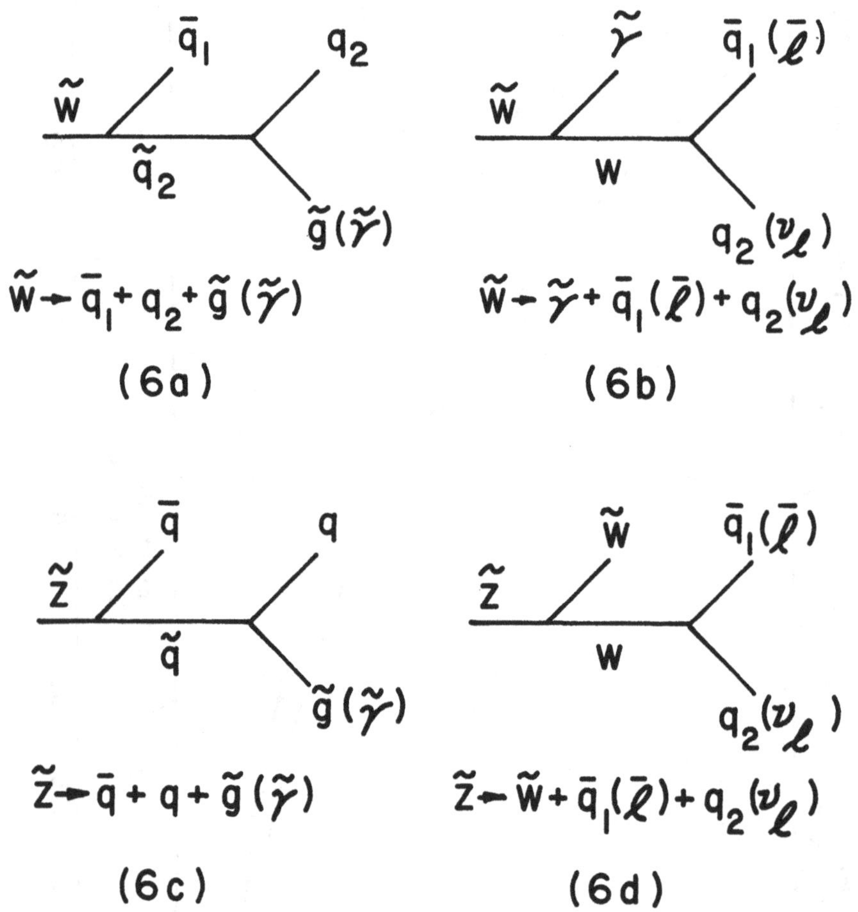
$\tilde{W}$
$\bar{q}_1$
q_2
$\tilde{q}_2$
$\tilde{g}(\tilde{\gamma})$
$\tilde{W} \to \bar{q}_1 + q_2 + \tilde{g}\,(\tilde{\gamma})$
(6a)
$\tilde{W}$
$\tilde{\gamma}$
$\bar{q}_1\,(\bar{\ell})$
W
$q_2\,(\nu_\ell)$
$\tilde{W} \to \tilde{\gamma} + \bar{q}_1\,(\bar{\ell}) + q_2(\nu_\ell)$
(6b)
$\tilde{Z}$
$\bar{q}$
q
$\tilde{q}$
$\tilde{g}\,(\tilde{\gamma})$
$\tilde{Z} \to \bar{q} + q + \tilde{g}\,(\tilde{\gamma})$
(6c)
$\tilde{Z}$
$\tilde{W}$
$\bar{q}_1(\bar{\ell})$
W
$q_2(\nu_\ell)$
$\tilde{Z} \to \tilde{W} + \bar{q}_1(\bar{\ell}) + q_2(\nu_\ell)$
(6d)

Fig. (6.)

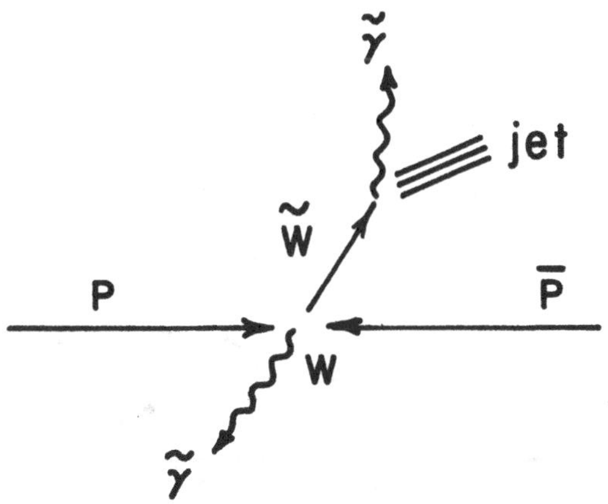

Fig. (7a): One jet UFO event in $W \rightarrow \tilde{W}\tilde{\gamma}$

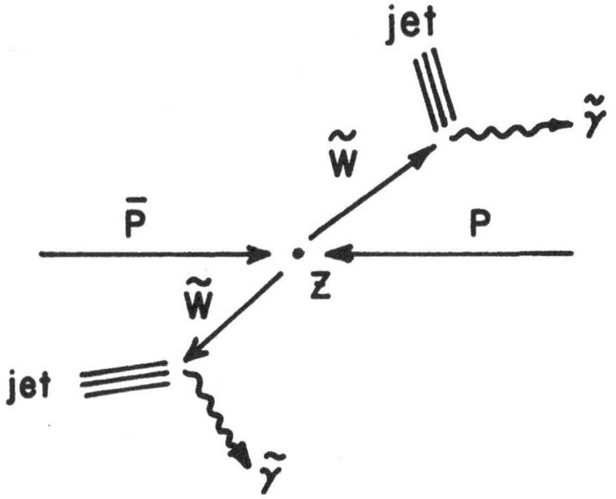

Fig. (7b): Two jet UFO event in $Z \rightarrow \tilde{W}+\tilde{W}$

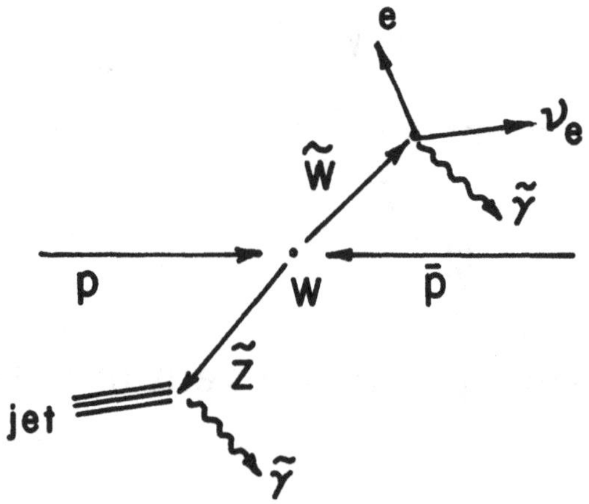

Fig. (8): Lepton-jet event in $W \to \tilde{W} + \tilde{Z}$